T0222159

General Game Playing

Synthesis Lectures on Artificial Intelligence and Machine Learning

Editors
Ronald J. Brachman, *Yahoo! Research*
William W. Cohen, *Carnegie Mellon University*
Peter Stone, *University of Texas at Austin*

General Game Playing
Michael Genesereth and Michael Thielscher
2014

Judgment Aggregation: A Primer
Davide Grossi and Gabriella Pigozzi
2014

An Introduction to Constraint-Based Temporal Reasoning
Roman Barták, Robert A. Morris, and K. Brent Venable
2014

Exact Algorithms for Probabilistic and Deterministic Graphical Models
Rina Dechter
2013

A Concise Introduction to Models and Methods for Automated Planning
Hector Geffner and Blai Bonet
2013

Essential Principles for Autonomous Robotics
Henry Hexmoor
2013

Case-Based Reasoning: A Concise Introduction
Beatriz López
2013

iv

Answer Set Solving in Practice
Martin Gebser, Roland Kaminski, Benjamin Kaufmann, and Torsten Schaub
2012

Planning with Markov Decision Processes: An AI Perspective
Mausam and Andrey Kolobov
2012

Active Learning
Burr Settles
2012

Computational Aspects of Cooperative Game Theory
Georgios Chalkiadakis, Edith Elkind, and Michael Wooldridge
2011

Representations and Techniques for 3D Object Recognition and Scene Interpretation
Derek Hoiem and Silvio Savarese
2011

A Short Introduction to Preferences: Between Artificial Intelligence and Social Choice
Francesca Rossi, Kristen Brent Venable, and Toby Walsh
2011

Human Computation
Edith Law and Luis von Ahn
2011

Trading Agents
Michael P. Wellman
2011

Visual Object Recognition
Kristen Grauman and Bastian Leibe
2011

Learning with Support Vector Machines
Colin Campbell and Yiming Ying
2011

Algorithms for Reinforcement Learning
Csaba Szepesvári
2010

Data Integration: The Relational Logic Approach
Michael Genesereth
2010

Markov Logic: An Interface Layer for Artificial Intelligence
Pedro Domingos and Daniel Lowd
2009

Introduction to Semi-Supervised Learning
XiaojinZhu and Andrew B.Goldberg
2009

Action Programming Languages
Michael Thielscher
2008

Representation Discovery using Harmonic Analysis
Sridhar Mahadevan
2008

Essentials of Game Theory: A Concise Multidisciplinary Introduction
Kevin Leyton-Brown and Yoav Shoham
2008

A Concise Introduction to Multiagent Systems and Distributed Artificial Intelligence
Nikos Vlassis
2007

Intelligent Autonomous Robotics: A Robot Soccer Case Study
Peter Stone
2007

General Game Playing

Michael Genesereth and Michael Thielscher

ISBN: 978-3-031-00441-4   paperback
ISBN: 978-3-031-01569-4   ebook

DOI 10.2200/S00564ED1V01Y201311AIM024

A Publication in the Springer series
SYNTHESIS LECTURES ON ARTIFICIAL INTELLIGENCE AND MACHINE LEARNING

Lecture #24
Series Editors: Ronald J. Brachman, *Yahoo Research*
                William W. Cohen, *Carnegie Mellon University*
                Peter Stone, *University of Texas at Austin*
Series ISSN
Synthesis Lectures on Artificial Intelligence and Machine Learning
Print 1939-4608   Electronic 1939-4616

General Game Playing

Michael Genesereth
Stanford University

Michael Thielscher
University of New South Wales

SYNTHESIS LECTURES ON ARTIFICIAL INTELLIGENCE AND MACHINE LEARNING #24

ABSTRACT

General game players are computer systems able to play strategy games based solely on formal game descriptions supplied at "runtime". (In other words, they don't know the rules until the game starts.) Unlike specialized game players, such as Deep Blue, general game players cannot rely on algorithms designed in advance for specific games; they must discover such algorithms themselves. General game playing expertise depends on intelligence on the part of the game player and not just intelligence of the programmer of the game player.

GGP is an interesting application in its own right. It is intellectually engaging and more than a little fun. But it is much more than that. It provides a theoretical framework for modeling discrete dynamic systems and defining rationality in a way that takes into account problem representation and complexities like incompleteness of information and resource bounds. It has practical applications in areas where these features are important, e.g., in business and law. More fundamentally, it raises questions about the nature of intelligence and serves as a laboratory in which to evaluate competing approaches to artificial intelligence.

This book is an elementary introduction to General Game Playing (GGP). (1) It presents the theory of General Game Playing and leading GGP technologies. (2) It shows how to create GGP programs capable of competing against other programs and humans. (3) It offers a glimpse of some of the real-world applications of General Game Playing.

KEYWORDS

general game playing, artificial intelligence, logic programming, computational logic, intelligent agents, knowledge representation

Contents

Preface . . . . **xv**

1 Introduction **1**
   1.1   Introduction 1
   1.2   Games 2
   1.3   Game Description 3
   1.4   Game Management 7
   1.5   Game Playing 8
   1.6   Discussion 9

2 Game Description **13**
   2.1   Introduction 13
   2.2   Logic Programs 13
   2.3   Game Model 17
   2.4   Game Description Language 20
   2.5   Game Description Example 21
   2.6   Game Simulation Example 24
   2.7   Game Requirements 27
   2.8   Prefix GDL 27

3 Game Management **31**
   3.1   Introduction 31
   3.2   Game Management 31
   3.3   Game Communication Language 32
   3.4   Game Play 33

4 Game Playing **37**
   4.1   Introduction 37
   4.2   Infrastructure 37
   4.3   Creating a Legal Player 38
   4.4   Creating a Random Player 41

5 Small Single-Player Games **43**

5.1 Introduction 43

5.2 8-Puzzle 43

5.3 Compulsive Deliberation 45

5.4 Sequential Planning 46

6 Small Multiple-Player Games **51**

6.1 Introduction 51

6.2 Minimax 51

6.3 Bounded Minimax Search 56

6.4 Alpha-Beta Search 59

7 Heuristic Search **67**

7.1 Introduction 67

7.2 Depth-Limited Search 67

7.3 Fixed-Depth Heuristic Search 68

7.4 Variable Depth Heuristic Search 70

8 Probabilistic Search **75**

8.1 Introduction 75

8.2 Monte Carlo Search 75

8.3 Monte Carlo Tree Search 78

9 Propositional Nets **81**

9.1 Introduction 81

9.2 Propositional Nets 82

9.3 Games as Propositional Nets 85

10 General Game Playing With Propnets **89**

10.1 Introduction 89

10.2 Propositional Nets as Data Structures 89

10.3 Marking and Reading Propositional Nets 92

10.4 Computing Game Playing Basics 94

11 **Factoring** **97**

    11.1 Introduction 97

    11.2 Compound Games with Independent Subgames 98

    11.3 Compound Games with Interdependent Termination 101

    11.4 Compound Games with Interdependent Actions 102

    11.5 Conditional Independence 104

12 **Discovery of Heuristics** **107**

    12.1 Introduction 107

    12.2 Latches 107

    12.3 Inhibitors 108

    12.4 Dead State Removal 108

13 **Logic** **111**

    13.1 Introduction 111

    13.2 Unification 111

    13.3 Derivation Steps (without Negation) 114

    13.4 Derivations 115

    13.5 Derivation Tree Search 117

    13.6 Handling Negation 119

14 **Analyzing Games with Logic** **129**

    14.1 Introduction 129

    14.2 Computing Domains 129

    14.3 Reducing the Domains Further 133

    14.4 Instantiating Rules 137

    14.5 Analyzing the Structure of GDL Rules 139

    14.6 Rule Graphs 139

    14.7 Using Rule Graphs 142

        14.7.1 Determining the Equivalence of Game Descriptions 142

        14.7.2 Computing Symmetries 142

    14.8 Exercises 145

15 **Solving Single-Player Games with Logic** **151**

    15.1 Answer Set Programming 151

    15.2 Adding Time to GDL Rules 152

15.3 Solving Single-Player Games with Answer Set Programming 156

15.4 Systems for Answer Set Programming 158

15.5 Exercises 158

16 Discovering Heuristics with Logic **161**

16.1 Discovering Heuristics with Answer Set Programming 161

16.2 Goal Heuristics 162

16.3 Fuzzy Logic 163

16.4 Using the Goal Heuristics 166

16.5 Optimizations and Limitations 167

16.6 Exercises 168

17 Games with Incomplete Information **171**

17.1 Introduction 171

17.2 GDL-II 171

17.3 Blind Tic-Tac-Toe 172

17.4 Card Games and Others 177

17.5 GDL-II Game Management 178

17.6 Playing GDL-II Games: Hypothetical States 180

17.7 Sampling Complete States 181

17.8 Exercises 186

18 Games with Historical Constraints **189**

18.1 Introduction 189

18.2 System Definition Language 189

18.3 Example–Tic-Tac-Toe 190

18.4 Example–Chess 194

19 Incomplete Game Descriptions **199**

19.1 Introduction 199

19.2 Relational Logic 199

19.3 Incomplete Game Description Language 203

19.4 Buttons and Lights Revisited 203

19.5 Complete Description of Buttons and Lights 205

19.6 Incomplete Description of Buttons and Lights 208

19.7 Playing Buttons and Lights with an Incomplete Description 208

20 Advanced General Game Playing **211**

20.1 Introduction 211

20.2 Temporal General Game Playing 211

20.3 Inductive General Game Playing...........................211

20.4 Really General Game Playing 212

20.5 Enhanced General Game Playing 212

Authors' Biographies **213**

Preface

General game players are computer systems able to play strategy games based solely on formal game descriptions supplied at "runtime". (In other words, they don't know the rules until the game starts.) Unlike specialized game players, such as Deep Blue, general game players cannot rely on algorithms designed in advance for specific games; they must discover such algorithms themselves. General game playing expertise depends on intelligence on the part of the game player and not just intelligence of the programmer of the game player.

General Game Playing (GGP) is an interesting application in its own right. It is intellectually engaging and more than a little fun. But it is much more than that. It provides a theoretical framework for modeling discrete dynamic systems and for defining rationality in a way that takes into account problem representation and complexities like incompleteness of information and resource bounds. It has practical applications in areas where these features are important, e.g., in business and law. More fundamentally, it raises questions about the nature of intelligence and serves as a laboratory in which to evaluate competing approaches to artificial intelligence.

This book is an elementary introduction to General Game Playing. (1) It presents the theory of GGP and leading GGP technologies. (2) It shows how to create GGP programs capable of competing against other programs and humans. (3) It offers a glimpse of some of the real world applications of General Game Playing.

Although the book is elementary, it does assume some basic background. First of all, readers should be familiar with Symbolic Logic. Game descriptions are written in the language of Symbolic Logic, and it helps to be able to read and write such descriptions. Second, readers should be familiar with the concepts of computer programming. At the very least, they should be able to read and understand program fragments written in modern programming languages. We use Javascript in all of our examples. Javascript is fairly simple. If readers are familiar with languages like Java and C, they should be able to read Javascript without any further training.

Before getting started, we want to acknowledge the contributions of various people. First of all, there are the various students who over the years helped to craft the course, notably Nathaniel Love, David Haley, Eric Schkufza, Evan Cox, Alex Landau, Peter Pham, Mirela Spasova, and Bertrand Decoster. Special mention goes to Sam Schreiber for maintaining the GGP configurable player and the Java code base used by many students. He is also the creator and maintainer or ggp.org, a website for all things GGP.

And thanks as well to the students who over the years have had to endure early versions of this material, in many cases helping to get it right by suffering through experiments that were not always successful. It is a testament to the intelligence of these students that they seem to have learned the material despite multiple bumbling mistakes on our part. Their patience and constructive comments were invaluable in helping us to understand what works and what does not.

Michael Genesereth and Michael Thielscher
March 2014

CHAPTER 1

Introduction

1.1 INTRODUCTION

Games of strategy, such as Chess, couple intellectual activity with competition. We can exercise and improve our intellectual skills by playing such games. The competition adds excitement and allows us to compare our skills to those of others. The same motivation accounts for interest in Computer Game Playing as a testbed for Artificial Intelligence. Programs that think better should be able to win more games, and so we can use competitions as an evaluation technique for intelligent systems.

Unfortunately, building programs to play specific games has limited value in AI. (1) To begin with, specialized game players are very narrow. They can be good at one game but not another. Deep Blue may have beaten the world Chess champion, but it has no clue how to play checkers. (2) A second problem with specialized game playing systems is that they do only part of the work. Most of the interesting analysis and design is done in advance by their programmers. The systems themselves might as well be tele-operated.

All is not lost. The idea of game playing can be used to good effect to inspire and evaluate good work in Artificial Intelligence, but it requires moving more of the design work to the computer itself. This can be done by focussing attention on General Game Playing.

General game players are systems able to accept descriptions of arbitrary games at runtime and able to use such descriptions to play those games effectively without human intervention. In other words, they do not know the rules until the games start.

Unlike specialized game players, such as Deep Blue, general game players cannot rely on algorithms designed in advance for specific games. General game playing expertise must depend on intelligence on the part of the game player and not just intelligence of the programmer of the game player. In order to perform well, general game players must incorporate results from various disciplines, such as knowledge representation, reasoning, and rational decision making; and these capabilities have to work together in a synergistic fashion.

Moreover, unlike specialized game players, general game players must be able to play different kinds of games. They should be able to play simple games (like Tic-Tac-Toe) and complex games (like Chess), games in static or dynamic worlds, games with complete and partial information, games with varying numbers of players, with simultaneous or alternating play, with or without communication among the players, and so forth.

1.2   GAMES

Despite the variety of games treated in General Game Playing, all games share a common abstract structure. Each game takes place in an environment with finitely many states, with one distinguished initial state and one or more terminal states. In addition, each game has a fixed, finite number of players; each player has finitely many possible actions in any game state, and each state has an associated goal value for each player. The dynamic model for general games is synchronous update: all players move on all steps (although some moves could be "no-ops"), and the environment updates only in response to the moves taken by the players.

Given this common structure, we can think of a game as a state graph, like the one shown in Figure 1.1. In this case, we have a game with one player, with eight states (named $s_1, \ldots, s_8$), with one initial state (s_1), with two terminal states (s_4 and s_8). The numbers associated with each state indicate the values of those states. The arcs in this graph capture the transition function for the game. For example, if the game is in state s_1 and the player does action a, the game will move to state s_2. If the player does action b, the game will move to state s_5.

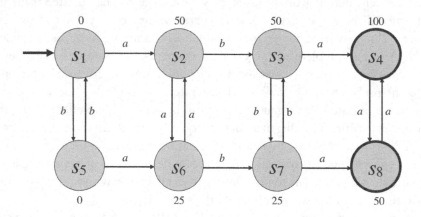

Figure 1.1: State graph for a single-player game.

In the case of multiple players with simultaneous moves, the arcs become multi-arcs, with one arc for each combination of the players' actions. Figure 1.2 gives an example of a simultaneous move game with two players. If in state s_1 both players perform action a, we follow the arc labeled a/a. If the first player does b and the second player does a, we follow the b/a arc. We also have different goals for the different players. For example, in state s_4, player 1 gets 100 points whereas player 2 get 0 points; and, in state s_8, the situation is reversed.

This conceptualization of games is an alternative to the traditional extensive normal form definition of games in game theory. While extensive normal form is more appropriate for certain types of analysis, the state-based representation has advantages in General Game Playing.

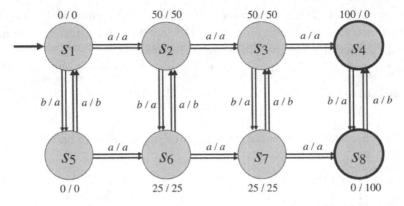

Figure 1.2: State graph for a two-player game.

In extensive normal form, a game is modeled as a game tree. In a game tree, each node is linked to successors by arcs corresponding to the actions legal in the corresponding game state. While different nodes often correspond to different states, it is possible for different nodes to correspond to the same game state. (This happens when different sequences of actions lead to the same state.)

In state-based representation, a game is modeled as a graph in which nodes are in 1-1 correspondence with states. Moreover, all players' moves are synchronous. (With extensions, extensive normal form can also represent simultaneous move games but with some added cost of complexity.) Additionally, state-based representation makes it possible to describe games more compactly, and it makes it easier for players to play games efficiently.

1.3   GAME DESCRIPTION

Since all of the games that we are considering are finite, it is possible, in principle, to describe such games in the form of state graphs. Unfortunately, such explicit representations are not practical in all cases. Even though the numbers of states and actions are finite, these sets can be extremely large; and the corresponding graphs can be larger still. For example, in Chess, there are thousands of possible moves and more than 10^{30} states.

In the vast majority of games, states and actions have composite structure that allows us to define a large number of states and actions in terms of a smaller number of more fundamental entities. In Chess, for example, states are not monolithic; they can be conceptualized in terms of pieces, squares, rows and columns and diagonals, and so forth.

By exploiting this structure, it is possible to encode games in a form that is more compact than direct representation. *Game Description Language* (GDL) supports this by relying on a conceptualization of game states as databases and by relying on logic to define the notions of legality and so forth.

As an example of GDL, let us look at the rules for the game of Tic-Tac-Toe. Note that this example is intended as a brief glimpse at GDL, not a rigorous introduction to the language. Full details of the language are given in the next chapter.

We begin with an enumeration of roles. There are two players: white and black.

```
role(white)
role(black)
```

Next, we characterize the initial state. In this case, all cells are blank.

```
init(cell(1,1,b))
init(cell(1,2,b))
init(cell(1,3,b))
init(cell(2,1,b))
init(cell(2,2,b))
init(cell(2,3,b))
init(cell(3,1,b))
init(cell(3,2,b))
init(cell(3,3,b))
init(control(white))
```

Next, we define legality. A player may mark a cell if that cell is blank and it has control. Otherwise, so long as there is a blank cell, the only legal action is noop, i.e., non-action. In GDL, symbols that begin with capital letters are variables, while symbols that begin with lower case letters are constants. The :- operator is read as "if" - the expression to its left is true if the expressions that follow it are true.

```
legal(W,mark(X,Y)) :-
  true(cell(X,Y,b))  &
  true(control(W))

legal(white,noop) :-
  true(cell(X,Y,b)) &
  true(control(black))

legal(black,noop) :-
  true(cell(X,Y,b)) &
  true(control(white))
```

Next, we look at the update rules for the game. A cell is marked with an x or an o if the corresponding player marks that cell. If a cell contains a mark, it retains that mark on the subsequent state. If a cell is blank and is not marked on that move, then it remains blank. Finally, control alternates on each play.

```
next(cell(M,N,x)) :-
  does(white,mark(M,N)) &
  true(cell(M,N,b))

next(cell(M,N,o)) :-
  does(black,mark(M,N)) &
  true(cell(M,N,b))

next(cell(M,N,W)) :-
  true(cell(M,N,W)) &
  distinct(W,b)

next(cell(M,N,b)) :-
  does(W,mark(J,K)) &
  true(cell(M,N,b)) &
  distinct(M,J)

next(cell(M,N,b)) :-
  does(W,mark(J,K)) &
  true(cell(M,N,b)) &
  distinct(N,K)

next(control(white)) :-
  true(control(black))

next(control(black)) :-
  true(control(white))
```

White gets a score of 100 points if there is a line of x's and no line of o's. It gets 0 points if there is a line of o's and no line of x's. Otherwise, it gets 50 points. The rewards for Black are analogous. The line relation is defined below.

```
goal(white,100) :- line(x) & ~line(o)
goal(white,50) :- ~line(x) & ~line(o)
goal(white,0) :- ~line(x) & line(o)

goal(black,100) :- ~line(x) & line(o)
goal(black,50) :- ~line(x) & ~line(o)
goal(black,0) :- line(x) & ~line(o)
```

Supporting concepts. A line is a row of marks of the same type or a column or a diagonal. A row of marks mean that's there three marks all with the same first coordinate. The column and diagonal relations are defined analogously.

```
line(Z) :- row(M,Z)
line(Z) :- column(M,Z)
line(Z) :- diagonal(Z)

row(M,Z) :-
  true(cell(M,1,Z)) &
  true(cell(M,2,Z)) &
  true(cell(M,3,Z))

column(N,Z) :-
  true(cell(1,N,Z)) &
  true(cell(2,N,Z)) &
  true(cell(3,N,Z))

diagonal(Z) :-
  true(cell(1,1,Z)) &
  true(cell(2,2,Z)) &
  true(cell(3,3,Z))

diagonal(Z) :-
  true(cell(1,3,Z)) &
  true(cell(2,2,Z)) &
  true(cell(3,1,Z))
```

Termination. A game terminates whenever there is a line of marks of the same type or if the game is no longer open, i.e., there are no blank cells.

```
terminal :- line(x)
terminal :- line(o)
terminal :- ~open

open :- true(cell(M,N,b))
```

Note that, under the full information assumption, any of these relations can be assumed to be false if it is not provably true. Thus, we have complete definitions for the relations legal, next, goal, and terminal in terms of true and does. The true relation starts out identical to init and on each step is changed to correspond to the extension of the next relation on that step.

Although GDL is designed for use in defining complete information games, it can be extended to partial information games relatively easily. Unfortunately, the resulting descriptions are more verbose and more expensive to process.

1.4   GAME MANAGEMENT

Game Management is the process of administering a game in a general game playing setting. More properly, it should be called match management, as the issue is how to manage individual matches of games, not the games themselves. However, everyone seems to use the phrase Game Management, and so we are stuck with it.

The process of running a game goes as follows. Upon receiving a request to run a match, the Game Manager's first sends a start message to each player to initiate the match. Once game play begins, it sends play messages to each player to get their plays and simulates the results. This part of the process repeats until the game is over. The Manager then sends stop messages to each player.

The start message lists the name of the match, the role the player is to assume (e.g., white or black in Chess), a formal description of the associated game (in GDL), and the start clock and play clock associated with the match. The start clock determines how much time remains before play begins. The play clock determines how much time each player has to make each move once play begins.

Upon receiving a start message, each player sets up its data structures and does whatever analysis it deems desirable in the time available. It then replies to the Game Manager that it is ready for play.

Having sent the start message, the game manager waits for replies from the players. Once it has received these replies OR once the start clock is exhausted, the Game Manager commences play.

On each step, the Game Manager sends a play message to each player. The message includes information about the actions of all players on the preceding step. (On the first step, the argument is nil.)

On receiving a play message, players spend their time trying to decide their moves. They must reply within the amount of time specified by the match's play clock.

The Game Manager waits for replies from the players. If a player does not respond before the play clock is exhausted, the Game Manager selects an arbitrary legal move. In any case, once all players reply or the play clock is exhausted, the Game manager takes the specified moves or the legal moves it has determined for the players and determines the next game state. It then evaluates the termination condition to see if the game is over. If the game is not over, the game manager sends the moves of the players to all players and the process repeats.

Once a game is determined to be over, the Game Manager sends a stop message to each player with information about the last moves made by all players. The stop message allows players to clean up any data structures for the match. The information about previous plays is supplied so that players with learning components can profit from their experience.

Having stopped all players, the Game Manager then computes the rewards for each player, stores this information together with the play history in its database, and ceases operation.

1.5   GAME PLAYING

Having a formal description of a game is one thing; being able to use that description to play the game effectively is something else. In this section, we examine some of the problems of building general game players and discuss strategies for dealing with these difficulties.

Let us start with automated reasoning. Since game descriptions are written in logic, it is obviously necessary for a game player to do some degree of automated reasoning.

There are various choices here. (1) A game player can use the game description interpretively throughout a game. (2) It can map the description to a different representation and use that interpretively. (3) It can use the description to devise a specialized program to play the game. This is effectively automatic programming.

The good news is that there are powerful reasoners for Logic freely available. The bad news is that such reasoners do not, in and of themselves, solve the real problems of general game playing, which are the same whatever representation for the game rules is used, viz. dealing with indeterminacy and resource bounds.

The simplest sort of game is one in which there is just one player and the number of states and actions is not too large. For such cases, traditional AI planning techniques are ideal. Depending on the shape of the search space, the player can search either forward or backward to find a sequence of actions / plays that convert the initial state into an acceptable goal state. Unfortunately, not all games are so simple.

To begin with, there is the *indeterminacy* that arises in games with multiple players. Recall that the succeeding state at each point in a game depends on the actions of all players, and remember that no player knows the actions of the other players in advance. Of course, in some cases, it is possible for a player to find sequences of actions guaranteed to achieve a goal state. However, this is quite rare. More often, it is necessary to create conditional plans in which a player's future actions are determined by its earlier actions and those of the other players. For such cases, more complex planning techniques are necessary.

Unfortunately, even this is not always sufficient. In some cases, there may be no guaranteed plan at all, not even a conditional plan. Tic-Tac-Toe is a game of this sort. Although it can be won, there is no guaranteed way to win in general. It is not really clear what to do in such situations. The key to winning is to move and hope that the moves of the other players put the game into a state from which a guaranteed win is possible. However, this strategy leaves open the question of which moves to make prior to arrival at such a state. One can fall back on probabilistic reasoning. However, this is not wholly satisfactory since there is no justifiable way of selecting a probability distribution for the actions of the other players. Another approach, of primary use in directly competitive games, is to make moves that create more search for the other players so that there is a chance that time limitations will cause those players to err.

Another complexity is the existence of resource bounds. In Tic-Tac-Toe, there are approximately 5000 distinct states. This is large but manageable. In Chess there are more than 10^{30} states. A state space of this size, being finite, is fully searchable in principle but not in practice. Moreover, the time limit on moves in most games means that players must select actions without knowing for sure whether they are any good.

In such cases, the usual approach is to conduct partial search of some sort, examining the game tree to a certain depth, evaluating the possible outcomes at that point, and choosing actions accordingly. Of course, this approach relies on the availability of an evaluation function for nonterminal states that is roughly monotonic in the actual probability of achieving a goal. While, for specific games, such as Chess, programmers are able to build in evaluation functions in advance, this is not possible for general game playing, since the structure of the game is not known in advance. Rather, the game player must analyze the game itself in order to find a useful evaluation function.

Another approach to dealing with size is abstraction. In some cases, it is possible to reformulate a state graph into a more abstract state graph with the property that any solution to the abstract problem has a solution when refined to the full state graph. In such cases, it may be possible to find a guaranteed solution or a good evaluation function for the full graph. Various researchers have proposed techniques along these lines, but more work is needed.

1.6   DISCUSSION

While general game playing is a topic with inherent interest, work in this area has practical value as well. The underlying technology can be used in a variety of other application areas, such as business process management and computational law. In fact, many games used in competitions are drawn from such areas.

General Game Playing is a setting within which AI is the essential technology. It certainly concentrates attention on the notion of specification-based systems (declarative systems, self-aware systems, and, by extension, reconfigurable systems, self-organizing systems, and so forth). Building systems of this sort dates from the early years of AI.

It was in 1958 that John McCarthy invented the concept of the "advice taker". The idea was simple. He wanted a machine that he could program by description. He would describe the intended environment and the desired goal, and the machine would use that information in determining its behavior. There would be no programming in the traditional sense. McCarthy presented his concept in a paper that has become a classic in the field of AI.

> *The main advantage we expect the advice taker to have is that its behavior will be improvable merely by making statements to it, telling it about its environment and what is wanted from it. To make these statements will require little, if any, knowledge of the program or the previous knowledge of the advice taker.*

An ambitious goal! But that was a time of high hopes and grand ambitions. The idea caught the imaginations of numerous subsequent researchers—notably Bob Kowalski, the high priest of logic programming, and Ed Feigenbaum, the inventor of knowledge engineering. In a paper written in 1974, Feigenbaum gave his most forceful statement of McCarthy's ideal.

> *The potential use of computers by people to accomplish tasks can be "one-dimensionalized" into a spectrum representing the nature of the instruction that must be given the computer to do its job. Call it the what-to-how spectrum. At one extreme of the spectrum, the user supplies his intelligence to instruct the machine with precision exactly how to do his job step-by-step. . . . At the other end of the spectrum is the user with his real problem. . . . He aspires to communicate what he wants done . . . without having to lay out in detail all necessary subgoals for adequate performance.*

Some have argued that the way to achieve intelligent behavior is through specialization. That may work so long as the assumptions one makes in building such systems are true. For general intelligence, however, general intellectual capabilities are needed, and such systems should be capable of performing well in a wide variety of tasks. To paraphrase the words of Robert Heinlein.

> *A human being should be able to change a diaper, plan an invasion, butcher a hog, conn a ship, design a building, write a sonnet, balance accounts, build a wall, set a bone, comfort the dying, take orders, give orders, cooperate, act alone, solve equations, analyze a new problem, pitch manure, program a computer, cook a tasty meal, fight efficiently, die gallantly. Specialization is for insects.*

It is our belief that general game playing offers an interesting application area within which general AI can be investigated.

PROBLEMS

Problem 1.1: Consider the following games. (Information about the games can be found on Wikipedia.)

Rubik's Cube	Go Fish
Minesweeper	Rock Paper Scissors
Diplomacy	Speed (card game)
World of Warcraft	Chess
Bughouse Chess	Stratego

For each of the following combinations of features, select a game from this list that manifests those features. (The classifications are not perfect in all cases.)

Players	Information	Moves	Communication
Single	Complete	N/A	N/A
Single	Partial	N/A	N/A
Multiple	Complete	Simultaneous	Yes
Multiple	Complete	Simultaneous	No
Multiple	Complete	Alternating	Yes
Multiple	Complete	Alternating	No
Multiple	Partial	Simultaneous	Yes
Multiple	Partial	Simultaneous	No
Multiple	Partial	Alternating	Yes
Multiple	Partial	Alternating	No

Problem 1.2: Connect to Gamemaster at using your favorite browser. Click the Games link and look at some of the games recorded there. Play a match of one of the games. The point of the exercise is to become familiar with the practice of learning and playing new games.

CHAPTER 2

Game Description

2.1 INTRODUCTION

The most significant characteristic of General Game Playing is that players do not know the rules of games before those games begin. Game rules are communicated at runtime, and the players must be able to read and understand the descriptions they are given in order to play legally and effectively.

In general, game playing, information about games is typically communicated to players in a formal language called *Game Description Language*, or *GDL*. This chapter is an introduction to GDL and the issues that arise in using it to describe games.

We start the chapter with a review of Logic Programs. After this, we introduce the game model underlying GDL, define the language, look at a sample game description, and look at the use of this description in simulating a match of the game. We then talk about additional features of games that ensure that they are interesting. Finally, we summarize the prefix syntax for GDL used in most GGP competitions.

2.2 LOGIC PROGRAMS

GDL is a logic programming language. It is similar to other logic programming languages, such as Prolog; but there are some important differences. (1) The semantics of GDL is purely declarative (there are no procedural constructs like assert, retract, and cut). (2) GDL has restrictions that assure that all questions of logical entailment are decidable. (3) There are some reserved words (described below), which tailor the language to the task of defining games.

Logic Programs are built up from four disjoint classes of symbols, viz. *object constants, function constants, relation constants*, and *variables*. In what follows, we write such symbols as strings of letters, digits, and a few non-alphanumeric characters (e.g., "_"). Constants must begin with a lower case letter or digit. Examples include a, b, 123, comp225, and barack_obama. Variables must begin with an uppercase letter. Examples include X, Y, Z, Age14, and so forth.

A *term* is either an object constant, a variable, or a functional term, i.e., an expression consisting of a function constant and *n* simpler terms. In what follows, we write functional terms in traditional mathematical notation—the function constant followed by its *arguments* enclosed in parentheses and separated by commas. For example, if f is a function constant, if a is an object constant, and if Y is a variable, then f(a,Y) is a term. Functional terms can be nested within other functional terms. For example, if f(a,Y) is a functional term, then so is f(f(a,Y),Y).

An *atom* is an expression formed from a relation constant and *n* terms. As with functional terms, we write atoms in traditional mathematical notation—the relation constant followed by its *arguments* enclosed in parentheses and separated by commas. For example, if r is a relation constant, if f is a function constant, if a is an object constant, and if Y is a variable, then r(a,Y) is a term and so is r(a,f(a,Y)). Although functional terms can be used within functional terms and within atoms, the reverse is not true—atoms cannot be nested inside of other atoms or inside of functional terms.

A *literal* is either an atom or a negation of an atom. A simple atom is called a *positive* literal, the negation of an atom is called a *negative* literal. In what follows, we write negative literals using the negation sign ~. For example, if p(a,b) is an atom, then ~p(a,b) denotes the negation of this atom.

A *rule* is an expression consisting of a distinguished atom, called the *head*, and a conjunction of zero or more literals, called the *body*. The literals in the body are called *subgoals*. In what follows, we write rules as in the example shown below. Here, q(X,Y) is the head; p(X,Y) & ~r(Y) is the body; and p(X,Y) and ~r(Y) are subgoals.

```
q(X,Y) :- p(X,Y) & ~r(Y)
```

A *logic program* is a finite set of atoms and rules of this form. In order to simplify our definitions and analysis, we occasionally talk about infinite sets of rules. While these sets are useful, they are not themselves logic programs.

Note: The following few paragraphs provide additional details about the syntax and semantics of logic programs. Readers who do not care about these technicalities at this point may want to jump ahead to the example of computing the minimal model of a logic program later in this section.

A rule in a logic program is *safe* if and only if every variable that appears in the head or in any negative literal in the body also appears in at least one positive literal in the body. A logic program is safe if and only if every rule in the program is safe.

The *dependency graph* for a logic program is a directed graph in which the nodes are the relations in the program and in which there is an arc from one node to another if and only if the former node appears in the body of a rule in which the latter node appears in the head. A program is *recursive* if and only if there is a cycle in the dependency graph.

A negation in a logic program is said to be *stratified* if and only if there is no recursive cycle in the dependency graph involving a negation. A logic program is stratified with respect to negation if and only if there are no unstratified negations.

The recursion in a set of rules is said to be *stratified* if and only if every variable in every subgoal relation (including the recursive relation) occurs in a subgoal involving a relation at a lower stratum, i.e., either it is not recursive or its recursive cycle does not include the relation in the head of the rule in which it occurs as a subgoal.

In GDL, we concentrate exclusively on logic programs that are both safe and stratified with respect to negation and recursion. While it is possible to extend the results here to programs that are not safe and stratified, such extensions are beyond the scope of this work.

The *Herbrand universe* for a logic program is the set of all terms that can be formed from the constants in the program's schema. Said another way, it is the set of all objects constants and all functional terms of the form $f(t_1, \ldots, t_n)$, where f is an n-ary function constant and $t_1, \ldots, t_n$ are elements of the Herbrand universe.

The *Herbrand base* for a logic program is the set of all atoms that can be formed from the constants in the program's schema. Said another way, it is the set of all sentences of the form $r(t_1, \ldots, t_n)$, where r is an n-ary relation constant and $t_1, \ldots, t_n$ are elements of the Herbrand universe.

A *dataset* for a logic program is an arbitrary subset of the Herbrand base for the program. A *model* of a logic program is a dataset that satisfies the program (as defined below).

An *instance* of a rule in a logic program is a rule in which all variables have been consistently replaced by terms from the program's Herbrand universe. *Consistent replacement* means that, if one occurrence of a variable is replaced by a given term, then all occurrences of that variable are replaced by the same term.

A dataset D *satisfies* a logic program P if and only if D satisfies every ground instance of every sentence in P. The notion of satisfaction is defined recursively. An interpretation D satisfies a ground atom p if and only if p is in D. D satisfies a negative ground literal $\sim p$ if and only if p is *not* in D. D satisfies a ground rule $p: -p_1 \& \ldots \& p_n$ if and only if D satisfies p whenever it satisfies $p_1, \ldots, p_n$.

In general, a logic program can have more than one model, which means that there can be more than one way to satisfy the rules in the program. In order to eliminate ambiguity, we adopt the minimal model approach to logic program semantics, i.e., we define the meaning of a safe and stratified logic program to be its minimal model.

A model D of a logic program P is *minimal* if and only if no proper subset of D is a model for P. A logic program that does not contain any negations has one and only one minimal model. A logic program with negation may have more than one minimal model; however, if the program is stratified, then once again there is only one minimal model. In general, models can be infinitely large. However, if the program has stratified recursion, then it is guaranteed to be finite.

Computing the minimal model for a logic program is conceptually easy. We initialize our dataset to the ground atoms in the program. We then look at the rules in the program. If there is an instance of a rule whose body is satisfied by the atoms in our dataset, then we add the corresponding instance of the head to the dataset. This process continues until it reaches a fixpoint, i.e., there are no additional ground atoms added by any rule. It can be shown that this process computes the unique minimal model for every logic program so long as it is safe and stratified with respect to negation and recursion.

As an example, consider the following logic program. We have a few facts about the parent relation; we have a rule defining the grandparent relation in terms of parent; and we have two rules defining ancestors in terms of parents.

```
parent(art,bob)
parent(art,bud)
parent(bob,cal)
parent(bob,coe)
parent(cal,dan)

grandparent(X,Z) :- parent(X,Y) & parent(Y,Z)

ancestor(X,Y) :- parent(X,Y)
ancestor(X,Z) :- ancestor(X,Y) & ancestor(Y,Z)
```

We start the computation by initializing our dataset to the five facts about parent.

```
parent(art,bob)
parent(art,bud)
parent(bob,cal)
parent(bob,coe)
parent(cal,dan)
```

Looking at the grandparent rule and matching its subgoals to the data in our dataset in all possible ways, we see that we can add the following data.

```
grandparent(art,cal)
grandparent(art,coe)
grandparent(bob,dan)
```

Looking at the ancestor rule and matching its subgoals to the data in our dataset in all possible ways, we get the following data.

```
ancestor(art,bob)
ancestor(art,bud)
ancestor(bob,cal)
ancestor(bob,coe)
ancestor(cal,dan)
```

With these additions, we can derive the following additional data.

```
ancestor(art,cal)
ancestor(art,coe)
ancestor(bob,dan)
```

However, we are not done. Using the ancestor rule again, we can derive the following additional datum.

```
ancestor(art,dan)
```

At this point, none of the rules when applied to this collection of data produces any results that are not already in the set, and so the process terminates. The resulting collection of 17 facts is the minimal model.

Logic programs as just defined are *closed* in that they fix the meaning of all relations in the program. In *open* logic programs, some of the relations (the inputs) are undefined, and other relations (the outputs) are defined in terms of these. The same program can be used with different input relations, yielding different output relations in each case.

Formally, an *open program* is a logic program together with a partition of the relation constants into two types—*base relations* (also called *input relations*) and *view relations* (also called *output relations*). View relations can appear anywhere in the program, but base relations can appear only in the subgoals of rules, not in their heads.

The *input base* for an open logic program is the set of all atoms that can be formed from the base relations of the program and the entities in the program's domain. An *input model* is an arbitrary subset of its input base.

The *output base* for an open logic program is the set of all atoms that can be formed from the view relations of the program and the entities in the program's domain. An *output model* is an arbitrary subset of its output base.

Given an open logic program P and an input model D, we define the *overall model* corresponding to D to be the minimal model of $P \cup D$. The output model corresponding to D is the intersection of the overall model with the program's output base; in other words, it consists of those sentences in the overall model that mention the output relations.

Finally, we define the meaning of an open logic program to be a function that maps each input model for the program into the corresponding output model.

2.3   GAME MODEL

The GDL model of games starts with entities and relations. *Entities* represent objects presumed or hypothesized to exist in the game. *Relations* represent properties of those objects or relationships among them.

In our examples here, we refer to entities and relations using strings of letters, digits, and a few non-alphanumeric characters (e.g., "_"). For reasons described below, we prohibit strings beginning with upper case letters; all other combinations are acceptable. Examples include x, o, 123, and white_king.

The set of all entities that can be used in a game is called the *domain* of the game. The set of all relations in a game is called the *signature* of the game. In GDL, domains and signatures are always finite (albeit in some cases very, very large).

The *arity* of a relation is the number of objects involved in any instance of that relation. Arity is an inherent property of a relation and never changes.

A game *schema* consists of a domain, a signature, and an assignment of arities for each of the relations in the signature.

Given a game schema, we define a *proposition* to be a structure consisting of an n-ary relation from the signature and n objects from the domain. In what follows, we write propositions using traditional mathematical notation. For example, if r is a binary relation and a and b are entities, then r(a,b) is a proposition.

The *propositional base* for a game is the set of all propositions that can be formed from the relations and the entities in the game's schema. For a schema with entities a and b and relations p and q where p has arity 1 and q has arity 2, the propositional base is { p(a), p(b), q(a,a), q(a,b), q(b,a), q(b,b) }.

In GDL, propositions are usually partitioned into disjoint classes, viz. base propositions and effectory propositions (more commonly called *actions*). Base propositions represent conditions that are true in the state of a game, and effectory propositions represent actions performed by game players. (Later, in order to deal with partial information, we add sensory propositions (or *percepts*) to this partition. For now, we ignore percepts.)

Before proceeding, let's look at these concepts in the context of a specific game, viz. Tic-Tac-Toe. As entities, we include white and black (the roles of the game), 1, 2, 3 (indices of rows and columns on the Tic-Tac-Toe board), and x, o, b (meaning blank).

We use the ternary relation cell together with a row index and a column index and a mark to designate the proposition that the cell in the specified row and column contains the specified mark. For example, the datum cell(2,3,o) asserts that there is an o in the cell in row 2 and column 3. We use the unary relation control to say whose turn it is to mark a cell. For example, the proposition control(white) asserts that it is white's turn.

In Tic-Tac-Toe, there only two types of actions a player can perform—it can mark a cell or it can do nothing (which is what a player does when it is not his turn to mark a cell). The binary relation mark together with a row m and a column n designates the action of placing a mark in row m and column n. The mark placed there depends on who does the action. The 0-ary relation noop refers to the act of doing nothing.

A *state* of a game is an arbitrary subset of the game's base propositions. The propositions in a state are assumed to be true whenever the game in is that state, and all others are assumed to be false. For example, we can describe the Tic-Tac-Toe state shown below on the left with the set of propositions shown on the right.

    cell(1,1,x)
    cell(1,2,b)
    cell(1,3,b)
    cell(2,1,b)
    cell(2,2,o)
    cell(2,3,b)

```
cell(3,1,b)
cell(3,2,b)
cell(3,3,b)
control(white)
```

On each time step, each role in a game has one or more legal actions it can perform. The actions on the left below are the legal moves for white in the state shown above. In this state, black has only one legal action, viz. noop, as shown on the right.

```
mark(1,2)
mark(1,3)
mark(2,1)
mark(2,3)
mark(3,1)
mark(3,2)
mark(3,3)
```

```
noop
```

A *move* in a game corresponds to a list of actions, one for each role. As a move is performed, some base propositions become true and others become false, leading to a new set of true propositions and, consequently, a new state and possibly a new set of legal actions.

For every state and every move in a game, there is a unique next state. For example, starting in the state shown below on the left, if the players perform the actions shown on the arrow, the game will move to the state shown on the right.

```
cell(1,1,x)                 cell(1,1,x)
cell(1,2,b)                 cell(1,2,b)
cell(1,3,b)                 cell(1,3,b)
cell(2,1,b)                 cell(2,1,b)
cell(2,2,o)     mark(3,3)   cell(2,2,o)
cell(2,3,b)        ⇒        cell(2,3,b)
cell(3,1,b)       noop      cell(3,1,b)
cell(3,2,b)                 cell(3,2,b)
cell(3,3,b)                 cell(3,3,x)
control(white)              control(black)
```

Every game is assumed to have a unique initial state and one or more terminal states. Every state is also assumed to have a value for each player—the number of points the player gets if the game terminates in that state. For example, the state on the right above is worth 100 points to white and 0 points to black.

A game starts in the initial state. The players select and execute legal actions in that state. The game then moves on to the next state (based on the players' actions). This process repeats until the game enters a terminal state, at which point the game stops and the players are awarded the number of points associated with the terminal state.

2.4 GAME DESCRIPTION LANGUAGE

In GDL, we fix the meanings of some words in the language for all games (the *game-independent vocabulary*) while at the same time allowing game authors to use their own words for individual games (the *game-specific vocabulary*).

There are 101 game-independent object constants in GDL, viz. the base ten representations of the integers from 0–100, inclusive, i.e., $0, 1, 2, \ldots, 100$. These are included for use as utility values for game states, with 0 being low and 100 being high. GDL has no game-independent function constants. However, there are ten game-independent relation constants, viz. the ones shown below.

role(a) means that a is a role in the game.

input(r, a) means that a is a feasible action for role r.

base(p) means that p is a base proposition in the game.

init(p) means that the proposition p is true in the initial state.

true(p) means that the proposition p is true in the current state.

does(r, a) means that role r performs action a in the current state.

next(p) means that the proposition p is true in the next state.

legal(r, a) means it is legal for role r to play action a in the current state.

goal(r, n) means that player the current state has utility n for player r.

terminal means that the current state is a terminal state.

A GDL description is an open logic program with the following input and output relations. (1) A GDL game description must give complete definitions for **role**, **base**, **input**, **init**. (2) It must define **legal** and **goal** and **terminal** in terms of an input **true** relation. (3) It must define **next** in terms of input **true** and **does** relations. Since **does** and **true** are treated as inputs, there must not be any rules with either of these relations in the head.

We can describe these concepts abstractly. However, experience has shown that most people learn their meaning more easily through examples. In the next section, we look at a definition of one particular game, viz. Tic-Tac-Toe.

2.5   GAME DESCRIPTION EXAMPLE

We begin with an enumeration of roles. In this case, there are just two roles, here called x and o.

```
role(white)
role(black)
```

We can characterize the propositions of the game as shown below.

```
base(cell(M,N,x)) :- index(M) & index(N)
base(cell(M,N,o)) :- index(M) & index(N)
base(cell(M,N,b)) :- index(M) & index(N)

base(control(white))
base(control(black))
```

We can characterize the feasible actions for each role in similar fashion.

```
input(R,mark(M,N)) :- role(R) & index(M) & index(N)
input(R, noop) :- role(R)

index(1)
index(2)
index(3)
```

Next, we characterize the initial state by writing all relevant propositions that are true in the initial state. In this case, all cells are blank; and the x player has control.

```
init(cell(1,1,b))
init(cell(1,2,b))
init(cell(1,3,b))
init(cell(2,1,b))
init(cell(2,2,b))
init(cell(2,3,b))
init(cell(3,1,b))
init(cell(3,2,b))
init(cell(3,3,b))
init(control(white))
```

Next, we define legality. A player may mark a cell if that cell is blank and it has control. Otherwise, the only legal action is noop.

```
legal(W,mark(X,Y)) :-
  true(cell(X,Y,b)) &
```

```
      true(control(W))

  legal(white,noop) :-
    true(control(black))

  legal(black,noop) :-
    true(control(white))
```

Next, we look at the update rules for the game. A cell is marked with an x or an o if the appropriate player marks that cell. If a cell contains a mark, it retains that mark on the subsequent state. If a cell is blank and is not marked on that step, then it remains blank. Finally, control alternates on each play.

```
  next(cell(M,N,x)) :-
    does(white,mark(M,N)) &
    true(cell(M,N,b))

  next(cell(M,N,o)) :-
    does(black,mark(M,N)) &
    true(cell(M,N,b))

  next(cell(M,N,W)) :-
    true(cell(M,N,W)) &
    distinct(W,b)

  next(cell(M,N,b)) :-
    does(W,mark(J,K))
    true(cell(M,N,b)) &
    distinct(M,J)

  next(cell(M,N,b)) :-
    does(W,mark(J,K))
    true(cell(M,N,b)) &
    distinct(N,K)

  next(control(white)) :-
    true(control(black))

  next(control(black)) :-
    true(control(white))
```

Goals. The white player gets 100 points if there is a line of x marks and no line of o marks. If there are no lines of either sort, white gets 50 points. If there is a line of o marks and no line of x marks, then white gets 0 points. The rewards for black are analogous. The line relation is defined below.

```
goal(white,100) :- line(x) & ~line(o)
goal(white,50) :- ~line(x) & ~line(o)
goal(white,0) :- ~line(x) & line(o)

goal(black,100) :- ~line(x) & line(o)
goal(black,50) :- ~line(x) & ~line(o)
goal(black,0) :- line(x) & ~line(o)
```

Supporting concepts. A line is a row of marks of the same type or a column or a diagonal. A row of marks mean that's there three marks all with the same first coordinate. The column and diagonal relations are defined analogously.

```
line(Z) :- row(M,Z)
line(Z) :- column(M,Z)
line(Z) :- diagonal(Z)

row(M,Z) :-
   true(cell(M,1,Z)) &
   true(cell(M,2,Z)) &
   true(cell(M,3,Z))

column(N,Z) :-
   true(cell(1,N,Z)) &
   true(cell(2,N,Z)) &
   true(cell(3,N,Z))

diagonal(Z) :-
   true(cell(1,1,Z)) &
   true(cell(2,2,Z)) &
   true(cell(3,3,Z)) &

diagonal(Z) :-
   true(cell(1,3,Z)) &
   true(cell(2,2,Z)) &
   true(cell(3,1,Z)) &
```

Termination. A game terminates whenever either player has a line of marks of the appropriate type or if the board is not open, i.e., there are no cells containing blanks.

```
terminal :- line(W)
terminal :- ~open

open :- true(cell(M,N,b))
```

2.6   GAME SIMULATION EXAMPLE

As an exercise in logic programming and GDL, let's look at the outputs of the ruleset defined in the preceding section at various points during an instance of the game.

To start, we can use the ruleset to compute the roles of the game. This is simple in the case of Tic-Tac-Toe, as they are contained explicitly in the ruleset.

```
role(white)
role(black)
```

Similarly, we can compute the possible propositions. Remember that this gives a list of all such propositions; only a subset will be true in any particular state.

```
base(cell(1,1,x))    base(cell(1,1,o))    base(cell(1,1,b))
base(cell(1,2,x))    base(cell(1,2,o))    base(cell(1,2,b))
base(cell(1,3,x))    base(cell(1,3,o))    base(cell(1,3,b))
base(cell(2,1,x))    base(cell(2,1,o))    base(cell(2,1,b))
base(cell(2,2,x))    base(cell(2,2,o))    base(cell(2,2,b))
base(cell(2,3,x))    base(cell(2,3,o))    base(cell(2,3,b))
base(cell(3,1,x))    base(cell(3,1,o))    base(cell(3,1,b))
base(cell(3,2,x))    base(cell(3,2,o))    base(cell(3,2,b))
base(cell(3,3,x))    base(cell(3,3,o))    base(cell(3,3,b))
base(control(white))
base(control(black))
```

We can also compute the relevant actions of the game. The extension of the input relation in this case consists of the 20 sentences shown below.

```
input(white,mark(1,1))
input(white,mark(1,2))
input(white,mark(1,3))
input(white,mark(2,1))
input(white,mark(2,2))
input(white,mark(2,3))
input(white,mark(3,1))
```

```
input(white,mark(3,2))
input(white,mark(3,3))
input(white,noop)

input(black,mark(1,1))
input(black,mark(1,2))
input(black,mark(1,3))
input(black,mark(2,1))
input(black,mark(2,2))
input(black,mark(2,3))
input(black,mark(3,1))
input(black,mark(3,2))
input(black,mark(3,3))
input(black,noop)
```

The first step in playing or simulating a game is to compute the initial state. We can do this by computing the init relation. As with roles, this is easy in this case, since the initial conditions are explicitly listed in the program.

```
init(cell(1,1,b))
init(cell(1,2,b))
init(cell(1,3,b))
init(cell(2,1,b))
init(cell(2,2,b))
init(cell(2,3,b))
init(cell(3,1,b))
init(cell(3,2,b))
init(cell(3,3,b))
init(control(white))
```

Once we have these conditions, we can turn them into a state description for the first step by asserting that each initial condition is true.

```
true(cell(1,1,b))
true(cell(1,2,b))
true(cell(1,3,b))
true(cell(2,1,b))
true(cell(2,2,b))
true(cell(2,3,b))
true(cell(3,1,b))
true(cell(3,2,b))
true(cell(3,3,b))
```

```
true(control(white))
```

Taking this input data and the logic program, we can check whether the state is terminal. In this case, it is not.

We can also compute the goal values of the state; but, since the state is non-terminal, there is not much point in doing that. However, the description does give us the following values.

```
goal(white,50)
goal(black,50)
```

More interestingly, using this state description and the logic program, we can compute legal actions in this state (see below). The x player has nine possible actions (all marking actions), and the o player has just one (noop).

```
legal(white,mark(1,1))
legal(white,mark(1,2))
legal(white,mark(1,3))
legal(white,mark(2,1))
legal(white,mark(2,2))
legal(white,mark(2,3))
legal(white,mark(3,1))
legal(white,mark(3,2))
legal(white,mark(3,3))
legal(black,noop)
```

Let's suppose that the x player chooses the first legal action and the o player chooses its sole legal action. This gives us the following dataset for does.

```
does(white,mark(1,1))
does(black,noop)
```

Now, combing this dataset with the state description above and the logic program, we can compute what must be true in the next state.

```
next(cell(1,1,x))
next(cell(1,2,b))
next(cell(1,3,b))
next(cell(2,1,b))
next(cell(2,2,b))
next(cell(2,3,b))
next(cell(3,1,b))
next(cell(3,2,b))
next(cell(3,3,b))
next(control(black))
```

To produce a description for the resulting state, we substitute `true` for `next` in each of these sentences and repeat the process. This continues until we encounter a state that is terminal, at which point we can compute the goals of the players in a similar manner.

2.7   GAME REQUIREMENTS

The definitions in Section 2.4 constrain GDL to game descriptions from which it is possible to compute the legal actions of all players for each state and from which it is possible to compute the next state for each state from the actions of all players. However, there are additional constraints that limit the scope of GDL to avoid problematic games.

Termination. A game description in GDL *terminates* if all infinite sequences of legal moves from the initial state of the game reach a terminal state after a finite number of steps.

Playability. A game description in GDL is *playable* if and only if every role has at least one legal move in every non-terminal state reachable from the initial state.

Winnability. A game description in GDL is *strongly winnable* if and only if, for some role, there is a sequence of individual actions of that role that leads to a terminal state of the game where that role's goal value is maximal no matter what the other players do. A game description in GDL is *weakly winnable* if and only if, for every role, there is a sequence of *joint* actions of all roles that leads to a terminal state where that role's goal value is maximal.

Well-formedness. A game description in GDL is *well-formed* if it terminates and is both playable and weakly winnable.

In general, game playing, all well-formed single player games should be strongly winnable. Clearly, it is possible to generate game descriptions in GDL which are not well formed. Checking game descriptions to see if they are well formed can certainly be done in general by using brute-force methods (exploring the entire game tree); and, for some games, faster algorithms may exist. Game descriptions used in GGP competitions are always well-formed. However, in this book, we occasionally look at games that are not well formed for reasons of simplicity or pedagogy.

2.8   PREFIX GDL

The version of GDL presented here uses traditional infix syntax. However, this is not the only version of the language. There is also a version that uses prefix syntax.

Although some general game playing environments support Infix GDL, it is not universal. On the other hand, all current systems support Prefix GDL. Fortunately, there is a direct relationship between the two syntaxes, and it is easy to convert between them. There are just a few issues to worry about.

The first issue is the spelling of constants and variables. Prefix GDL is case-independent, so we cannot use capital letters to distinguish the two. Constants are spelled the same in both versions; but, in prefix GDL, we distinguish variables by beginning with the character '?'. Thus, the constant a is the same in both languages while the variable X in Infix GDL is spelled ?x or ?X in Prefix GDL.

The second issue in mapping between the formats is syntax of expressions. In Prefix GDL, all expressions are lists of components separated by spaces and enclosed in parentheses. Also, logical operators are spelled out. The following tables illustrates the mapping.

Infix GDL	Prefix GDL
p(a,Y)	(p a ?y)
~p(a,Y)	(not (p a ?y))
p(a,Y) & p(Y,c)	(and (p a ?y) (p ?y c))
q(Y) :- p(a,Y) & p(Y,c)	(<= (q ?y) (and (p a ?y) (p ?y c)))
q(Y) :- p(a,Y) & p(Y,c)	(<= (q ?y) (p a ?y) (p ?y c))

Finally, just to be clear on this, in Prefix GDL white space (spaces, tabs, carriage returns, line feeds, and so forth) can appear anywhere other than in the middle of constants, variables, and operator names. Thus, there can be multiple spaces between the components of an expression; there can be spaces after the open parenthesis of an expression and before the operator or relation constant or function constant; and there can be spaces after the last component of an expression and the closing parenthesis.

PROBLEMS

Problem 2.1: Consider the game description shown below.

```
role(white)        next(p) :- does(white,a) ~true(p)
role(black)        next(p) :- ~does(white,a) & true(p)
                   next(q) :- does(white,b) & true(p)
base(p)            next(q) :- does(white,c) & true(r)
base(q)            next(q) :- ~does(white,b) ~does(white,c) & true(q)
base(r)            next(r) :- does(white,c) & true(q)
base(s)            next(r) :- ~does(white,c) & true(r)

action(a)          goal(white,100) :- terminal
action(b)          goal(white,0) :- ~terminal
action(c)          goal(black,100) :- terminal
action(d)          goal(black,0) :- ~terminal

init(s)            terminal :- true(p) & true(q) & true(r)

legal(white,a)
legal(white,b)
legal(white,c)
```

`legal(black,d)`

(a) How many roles are there?

(b) How many propositions are there?

(c) How many feasible actions are there?

(d) How many actions are legal for white in the initial state?

(e) How many propositions are true in the initial state?

(f) How many are true in the state that results from white performing action a and black performing action d in the initial state?

(g) What is the minimum number of steps this game can take to terminate?

Problem 2.2: More questions about the game in Problem 2.1.

(a) Does the game always terminate?

(b) Is the game playable?

(c) Is the game strongly winnable for white?

(d) Is the game weakly winnable for white?

(e) Is the game strongly winnable for black?

(f) Is the game weakly winnable for black?

Problem 2.3: For each of the following pairs of expressions, say whether the expression on the second line is a faithful translation of the expression on the first line into Prefix GDL.

(a) r(a,b) :- p(a) & q(b)
   (<= (r a b) (and (p a) (q b)))

(b) r(a,b) :- p(a) & q(b)
   (<= (r a b) (p a) (q b))

(c) r(x,y) :- p(x) & q(y)
   (<= (r ?x ?y) (p ?x) (q ?y))

(d) r(X,Y) :- p(X) & q(Y)
   (<= (r ?x ?y) (p ?x) (q ?y))

CHAPTER 3

Game Management

3.1 INTRODUCTION

This chapter is an overview of game management. More properly, it should be called match management, as the issue is how to manage individual matches of games, not the games themselves. We start with an overview of the General Game Playing ecosystem and the central role of the Game Manager. We then discuss the General Game Playing communication protocol. Finally, we see how it is used in a sample game.

3.2 GAME MANAGEMENT

A diagram of a typical general game playing ecosystem is shown below. At the center of the ecosystem is the game manager. The game manager maintains a database of game descriptions and match records, and it maintains some temporary state for matches while they are running. The game manager communicates with game players. It also provides a user interface for users who want to schedule matches, and it provides graphics for spectators watching matches in progress.

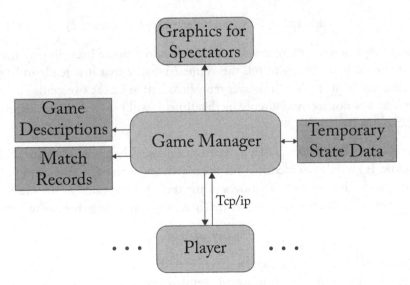

3.3 GAME COMMUNICATION LANGUAGE

Communication between the Game Manager and game players takes place through HTTP connections. The communication model assumes that each player is running on an Internet-connected host listening on a particular port. HTTP messages sent to players have the standard HTTP header, with content type `text/acl`.

In the current GGP communication language, there are five types of messages used for communication between the Game Manager and game players.

(1) An `info` message is used to confirm that a player is up and running. The general form is shown below.

$$info()$$

Upon receipt of an `info` message, a player is expected to return `available` if it is ready to receive messages. Otherwise, it should return `busy`. It may also return arbitrary information about itself as a sequence of pairs of type and value, e.g., `[[name,egghead],[status,available]]`.

(2) A `start` message is used to initialize an instance of a game. The general form of a `start` message is shown below. The message begins with the keyword `start`, and this is followed by five arguments, viz. a match identifier (a sequence of alphanumeric characters beginning with a lowercase letter), a role for the player to play (chosen from the roles mentioned in the game description), a list of game rules (written as sentences in GDL), a start clock (in seconds), and a play clock (also in seconds).

$$start(\textit{id, role, description, startclock, playclock})$$

Upon receipt of a `start` message, a player should prepare itself to play the match. Once it is done, it should reply `ready` to tell the Game Manager that it is ready to begin the match. The GGP protocol requires that the player reply before `startclock` seconds have elapsed. If the Game Manager has not received a reply by this time, it will proceed on the assumption that the player is ready.

(3) A `play` message is used to request a move from a player. The general form of the `play` message is shown below. It includes an identifier for the match and a record of the moves of all players on the preceding step. The order of the moves in the record is the same as the order of roles in the game description. On the first request, where there is no preceding move, the move field is `nil`.

$$play(\textit{id, move})$$

Upon receipt of a `play` message, a player uses the move information to update its state as necessary. It then computes its next move and returns that as answer. The GGP protocol requires that the player reply before `playclock` seconds have elapsed. If the Game Manager has not received a reply by this time, it substitutes an arbitrary legal move.

(4) A stop message is used to tell a player that a match has reached completion. The general form of the stop message is shown below.

$$\text{stop}(id, move)$$

Upon receipt of a stop message, a player can clean up after playing the match. The move is sent along in case the player wants to know the move that terminated the game. After finishing up, the player should return done.

(5) An abort message is used to tell a player that a match is terminating abnormally. It differs from a stop message in that the match need not be in a terminal state.

$$\text{abort}(id)$$

Upon receipt of an abort message, a player can eliminate any data structures and return to a ready state. Once it is finished, it should return done.

3.4 GAME PLAY

The process of running a match goes as follows. Upon receiving a request to run a match, the Game Manager first sends a start message to each player to initiate the match. Once game play begins, the manager sends play messages to each player to get their plays; and it then simulates the results. This part of the process repeats until the game is over. The Manager then sends a stop message to each player.

Here is a sample of messages for a quick game of Tic-Tac-Toe. The game manager initiates the match by sending a start message to all of the players, each with a different role. The players then respond with ready. They can respond immediately or they can wait until the start clock is exhausted before responding.

Game Manager to Player x:	start(m23,x,[role(x),role(o),...],10,10)
Game Manager to Player y:	start(m23,y,[role(x),role(o),...],10,10)
Player x to Game Manager:	ready
Player y to Game Manager:	ready

Play begins after all of the players have responded or after the start clock has expired, whichever comes first. The manager initiates play by sending a play message to all of the players. Since this is the first move and there are no previous moves, the move argument in the play message is nil. In this case, the first player responds with the action mark(1,1), one of its nine legal actions; and the second player responds with noop, its only legal action.

Game Manager to Player x:	play(m23,nil)
Game Manager to Player y:	play(m23,nil)
Player x to Game Manager:	mark(1,1)
Player y to Game Manager:	noop

The Game Manager checks that these actions are legal, simulates their effects, updates the state of the game, and then sends `play` messages to the players to solicit their next actions. The second argument in the `play` message this time is a list of the actions received in response to the preceding `play` message. On this step, the first player responds with `noop`, its only legal action; and the second player responds with `mark(1,2)`. This is a legal move, but it is not a wise move, as the game is now strongly winnable by the first player.

Game Manager to Player x:	`play(m23,[mark(1,1),noop])`
Game Manager to Player y:	`play(m23,[mark(1,1),noop])`
Player x to Game Manager:	`noop`
Player y to Game Manager:	`mark(1,2)`

Again, the Game Manager checks legality, simulates the move, updates its state, and sends a `play` message requesting the players' next actions. The first player takes advantage of the situation and plays `mark(2,2)` while the second player does `noop`.

Game Manager to Player x:	`play(m23,[noop,mark(1,2)])`
Game Manager to Player y:	`play(m23 [noop,mark(1,2)])`
Player x to Game Manager:	`mark(2 2)`
Player y to Game Manager:	`noop`

There is not much the second player can do in this situation to save itself. Instead of staving off the immediate loss, it plays `mark(1,3)`, while the first player does `noop`.

Game Manager to Player x:	`play(m23,[mark(2,2),noop])`
Game Manager to Player y:	`play(m23,[mark(2,2),noop])`
Player x to Game Manager:	`noop`
Player y to Game Manager:	`mark(1,3)`

The Game Manager again simulates, updates, and requests a move. In this case, the first player goes in for the kill, playing `mark(3,3)`.

Game Manager to Player x:	`play(m23,[noop,mark(1,3)])`
Game Manager to Player y:	`play(m23,[noop,mark(1,3)])`
Player x to Game Manager:	`mark(3,3)`
Player y to Game Manager:	`noop`

With this move, the game is over. As usual, in such cases, the Manager lets the players know by sending a suitable `stop` message. It then stores the results in its database for future reference and terminates.

Game Manager to Player x:	`stop(m23,[mark(3,3),noop])`
Game Manager to Player y:	`stop(m23,[mark(3,3),noop])`
Player x to Game Manager:	`done`
Player y to Game Manager:	`done`

Note that the Game Manager sends slightly different `start` messages to the different players. Everything is the same except for the role that each player is asked to play. In all other ways, the messages sent to the players are identical. In advanced versions of the General Game Playing protocol, this symmetry is broken. The Game Manager can send different game descriptions to different players. And it can tell different players different information in the `play` and `stop` messages.

.

PROBLEMS

Problem 3.1: Consider the game of Tic-Tac-Toe given in the preceding chapter. Assume that a Game Manager has sent `start` messages to the players of a match with name m23 and with the rules from the preceding chapter as game description; and assume that the players have just replied that they are ready to play. Which of the following is the correct message for the Manager to send to the players next?

(*a*)   `play()`
(*b*)   `play(m23)`
(*c*)   `play(m23,nil)`
(*d*)   `play(m23,noop)`
(*e*)   `play(m23,[mark(1,1),noop])`

Problem 3.2: Consider the game of Tic-Tac-Toe given in the preceding chapter. Assume that the game is in the state shown on the left below, and assume that the manager has just received the action `mark(2,2)` from the first player and the action `noop` from the second player. Which of the messages shown on the right is the correct message to send to the first player?

X	O	

(*a*)   `play()`
(*b*)   `play(m23)`
(*c*)   `play(m23,mark(2,2))`
(*d*)   `play(m23,noop)`
(*e*)   `play(m23,[mark(2,2),noop])`

Problem 3.3: Consider the game of Tic-Tac-Toe given in the preceding chapter. Assume that the game is in the state shown on the left below, and assume that the manager has just received the action mark(2,2) from the first player and the action noop from the second player. Which of the messages shown on the right is a correct message to send to the first player.

X	O	
	X	O

(*a*)   play(m23)

(*b*)   play(m23,[mark(2,2),noop])

(*c*)   play(m23,[mark(1,3),noop])

(*d*)   stop(m23,[mark(2,2),noop])

(*e*)   abort(m23)

CHAPTER 4

Game Playing

4.1 INTRODUCTION

The simplest sort of general game player is one that explores the game tree implicit in a game description. In this chapter, we talk about some infrastructure that frames the problem more precisely. We then consider a couple of search-free uses of this infrastructure, viz. legal players and random players. In Chapters 5–8, we look at complete search techniques (which are appropriate for small game trees) as well as incomplete search techniques (which are necessary for very large game trees). In Chapters 9–12, we examine some game playing techniques based on properties of states. Finally, in Chapters 13–16, we show ways that game descriptions can be used to deduce general properties of games without explicitly enumerating states or properties.

4.2 INFRASTRUCTURE

A game player is typically implemented as a web service. The service receives messages from a Game Manager and replies appropriately. Building a player means writing event handlers for the different types of messages in the GGP communication protocol.

To make things concrete, let us assume that we have at our disposal a code base that (1) includes a listener to call these event handlers on receipt of messages from a Game Manager and (2) includes subroutines for processing game descriptions and match data. Our job in building a player is to use the available subroutines to implement event handlers for the various GGP messages.

Once running, the listener enters a loop listening for messages from a Game Manager. Upon receipt of a message, the listener calls the appropriate handler. When the handler is finished, the listener sends the returned value to the Game Manager. The handlers called by the listener are listed below.

info()

start(*id*, *role*, *rules*, *startclock*, *playclock*)

play(*id*, *move*)

stop(*id*, *move*)

abort(*id*)

In order to facilitate the implementation of these message handlers, we assume that our code base contains definitions for the subroutines described below. There are subroutines for computing most of the components of a match.

findroles(*game*) - returns a sequence of roles.

findpropositions(*game*) - returns a sequence of propositions.

findactions(*role, game*) - returns a sequence of actions for a specified role.

findinits(*game*) - returns a sequence of all propositions that are true in the initial state.

findlegalx(*role, state, game*) - returns the first action that is legal for the specified role in the specified state.

findlegals(*role, state, game*) - returns a sequence of all actions that are legal for the specified role in the specified state.

findnext(*roles, move, state, game*) - returns a sequence of all propositions that are true in the state that results from the specified roles performing the specified move in the specified state.

findreward(*role, state, game*) - returns the goal value for the specified role in the specified state.

findterminalp(*state, game*) - returns a boolean indicating whether the specified state is terminal.

That's it. As mentioned above, our job is to use these subroutines to write the handlers called by the listener. In the remainder of this chapter, we look at a couple of simple approaches to doing this.

4.3 CREATING A LEGAL PLAYER

A *legal player* is the simplest form of game player. In each state, a legal player selects an action based solely on its legality, without consideration of the consequences. Typically, the choice of action is consistent—it selects the same action every time it finds itself in the same state. (In this way, a legal player differs from a random player, which selects different legal actions on different occasions.)

Legal play is not a particularly good general game playing strategy. However, it is a worthwhile exercise to build a legal player (and a random player) just to get familiar with the concepts described above and to have a basis of comparison for more intelligent players.

Using the basic subroutines provided in the GGP starter pack, building a legal player is very simple. We start by setting up some global variables to maintain information while a match is in progress. (Properly, we should create a data structure for each match; and we should attach these values to this data structure. However, we are striving for simplicity of implementation in these notes. This does not mean that you should do the same.)

```
var game;
var role;
var roles;
var state;
```

Next, we define a handler for each type of message. The info handler simply returns ready.

```
function info ()
{return 'ready'}
```

The start event handler assigns values to game and role based on the incoming start message; it initializes state; and it then returns ready, as required by the GGP protocol.

```
function start (id,player,rules,sc,pc)
 {game = rules;
  role = player;
  roles = findroles(game);
  state = findinits(game);
  return 'ready'}
```

The play event handler takes a match identifier and a move as arguments. It first uses the simulate subroutine to compute the current state. If the move is nil, then this is the initial state, and the player uses findinits to compute the state based on the initial conditions supplied in the game description. Otherwise, it uses findnexts to compute the state resulting from the preceding state and the actions supplied in the move. Once our player has the latest state, it uses findlegalx to compute a legal move.

```
function play (id,move)
 {state = simulate(move,state);
  return findlegalx(role,state,game)}

function simulate (move,state)
 {if (move == 'nil') {return state};
  return findnexts(roles,move,state,game)}
```

The stop event handler for our legal player does nothing. It ignores the inputs and simply returns done as required by the GGP protocol.

```
function stop (id,move)
{return 'done'}
```

Like the stop message handler, the abort message handler for our player also does nothing. It simply returns done.

```
function abort (id)
{return 'done'}
```

Just to be clear on how this works, let's work through a short Tic-Tac-Toe match. When a player is initialized, it sets up data structures to hold the game description, the role, and the state. These are initially empty.

Let's assume that our player receives a start message from a Game Manager of the sort shown below. The match identifier is m23. Our player is asked to be the x player. There are the usual axioms of Tic-Tac-Toe. The start clock and play clock are both 10 s.

```
start(m23, white, [role(white),role(black),...], 10, 10)
```

On receipt of this message, our listener calls the start handler. This sets the global variables accordingly. The returned value ready is then sent back to the Game Manager.

Once the Game Manager is ready, it sends a suitable play message to all players. See below. Here we have a request for each player to choose an action for match m23. The argument nil signifies that this is the first step of the match.

```
play(m23,nil)
```

On receipt of this message, our listener invokes the play handler with the arguments passed to it by the Game Manager. Since the move is nil, our player computes the current state by calling findinits on the game description. This results in the dataset shown below.

```
true(cell(1,1,b))
true(cell(1,2,b))
true(cell(1,3,b))
true(cell(2,1,b))
true(cell(2,2,b))
true(cell(2,3,b))
true(cell(3,1,b))
true(cell(3,2,b))
true(cell(3,3,b))
true(control(white))
```

Using this state, together with the role and game description associated with this match, our player then computes the first legal move, i.e., mark(1,1) and returns that as answer.

The Game Manager checks that the actions of all players are legal, simulates their effects and updates the state of the game, and then sends play messages to the players to solicit their next actions. In this case, our player would receive the message shown below.

$$play(m23, [mark(1,1), noop])$$

Again, our player invokes its play handler with the arguments passed to it by the Game Manager. This time, the move is not nil, and so our player uses findnexts to compute the next state. This results in the dataset shown below.

```
true(cell(1,1,x))
true(cell(1,2,b))
true(cell(1,3,b))
true(cell(2,1,b))
true(cell(2,2,b))
true(cell(2,3,b))
true(cell(3,1,b))
true(cell(3,2,b))
true(cell(3,3,b))
true(control(black))
```

Using this state, our player then computes the first legal move, its only legal move, viz. noop, and returns that as answer.

This process then repeats until the end of the game, at which point our player receives a message like the one shown below.

$$stop(m23, [mark(3,3), noop])$$

While some players are able to make use of the information in a stop message, our legal player simply ignores this information and returns done, terminating its activity on this match.

4.4   CREATING A RANDOM PLAYER

A *random player* is similar to a legal player in that it selects an action for a state based solely on its legality, without consideration of the consequences. A random player differs from a legal player in that it does not simply take the first legal move it finds but rather selects randomly from among the legal actions available in the state, usually choosing a different move on different occasions.

The implementation of a random player is almost identical to the implementation of a legal player. The only difference is in the `play` handler. In selecting an action, our player first computes all legal moves in the given state and then randomly selects from among these choices (using the `randomelement` subroutine). One way of writing the code for the play handler is shown below.

```
function play (id,move)
 {state = simulate(move,state);
  var actions=findlegals(role,state,game);
  return randomelement(actions)}
```

Random players are no smarter than legal players. However, they often appear more interesting because they are unpredictable. Also, they sometimes avoid traps that befall consistent players like legal, which can sometimes maneuver themselves into a corner and be unable to escape. They are also used as standards to show that general game players or specific methods perform better than chance.

A random player consumes slightly more compute time than a legal player, since it must compute all legal moves rather than just one. For most games, this is not a problem; but for games with a large number of possible actions, the difference can be noticeable.

CHAPTER 5

Small Single-Player Games

5.1 INTRODUCTION

We start our tour of general game playing by looking at single-player games. In the game-playing community, these are often called *puzzles* rather than games; and the process of solving such puzzles is often called *problem-solving* rather than game-playing.

Puzzles are simpler than multiple-player games because everything is under the control of the single player. The world is static, except when the player acts; and changes to the world are determined entirely by the current state and the actions of the player.

In this chapter, as in most of this book, we assume that the player has complete information about the game. We assume that it knows the initial state; it knows all of its legal actions in every state; it knows the effects of its actions in every state; for every state, it knows its reward; and, for every state, it knows whether or not it is terminal.

In this chapter, we also assume the games are small, i.e., they are small enough so that there is sufficient time for the player to search the entire game tree. This guarantees that the player can find optimal actions to perform. That said, as we shall see, it is sometimes possible to find optimal actions even without searching the entire game tree.

Despite these strong assumptions (just one player, complete information, and the availability of adequate time to search the game tree), the study of single player games is a good place to start our look at general game playing. First of all, many real-world problems can be cast as single-player games with these same restrictions. More importantly for us, as we shall see, the techniques we examine later can be viewed as more elaborate versions of the basic techniques introduced here.

We begin this chapter with an example of a single-player game that we use throughout the chapter. We then look at two different approaches to single-player game playing.

5.2 8-PUZZLE

The 8-puzzle is a sliding tile puzzle. The game board is a 3×3 square with numbered tiles in all but one of the cells. See the example shown below.

The state of the game is modified by sliding numbered tiles into the empty space from adjacent cells, thus moving the empty space to a new location. There are four possible moves—moving the empty space up, down, left, or right. Obviously, not all moves are possible in all states. The states shown on the left and right below illustrate the possible moves from the state shown in the center.

The ultimate object of the game is to place the tiles in order and position the empty square in the lower right cell. See below.

The game terminates after eight moves or when all of the tiles are in the right positions, whichever comes first. Partial credit is given for states that approximate the ultimate goal, with 10 points being allocated for each numbered tile in the correct position and 20 points being allocated for having the empty tile in the correct position. For example, the initial state shown above, the one in the middle, is worth 40 points; and the goal state is worth 100 points.

5.3   COMPULSIVE DELIBERATION

Compulsive Deliberation is a particularly simple approach to game playing. On each step, the player examines the then-current game tree to determine its best move for that step; and it makes this move. It repeats this process on the next step and so forth until the end of the game.

In pure compulsive deliberation, each step of the computation is independent of every other step. No data computed during any step is accessible on subsequent steps. The player treats each step as if it were a new game. This is obviously wasteful, but it does not hurt so long as there is enough time to do the repeated calculations. We start with this method because it is simple to understand and at the same time serves as a template for the more sophisticated, less wasteful methods to come.

The following procedure is a simple implementation of compulsive deliberation. On receipt of a play message, the player simulates the specified move as usual. It then computes all of its legal actions in this new state and iterates through these actions comparing the score of each to the best score found so far. If it ever finds an action that guarantees a reward of 100, it stops and returns that action. Otherwise, it retains the highest score and action and returns the corresponding action when it is done.

```
function play (id,move)
 {state = simulate(move,state);
  return bestmove(role,state)}

function bestmove (role, state)
 {var actions = findlegals(role,state,game);
  var action = actions[0];
  var score = 0;
  for (var i=0; i<actions.length; i++)
      {var result = maxscore(role,simulate([actions[i]]),state));
       if (result==100) {return actions[i]};
       if (result>score) {score = result; action = actions[i]}};
  return action}
```

Recall that every game has a unique reward for each player in each state. The reward is a perfect measure of utility for terminal states. Unfortunately, in general, the rewards for non-terminal states do not always correlate with the rewards for terminal states; and so they are not necessarily useful in defining the scores of non-terminal states.

State utility is a measure of utility that is relevant for both terminal states and non-terminal states. By definition, the utility of a state for a player is defined as being the best reward the player can guarantee for itself by any sequence of legal moves starting in the given state.

One way to determine the utility of a state is by computing all sequences of legal actions that lead from the given state to a terminal state and taking the maximum reward for the terminal state resulting from each such sequence.

The following procedure computes the same score in a slightly simpler manner. The procedure takes a player and a state as arguments and returns the corresponding utility as score. The procedure computes the score via a recursive exploration of the game tree. If the state supplied as argument is terminal, the output is just the player's reward for that state. Otherwise, the output is the maximum of the utilities of the states resulting from executing any of the player's legal actions in the given state.

```
function maxscore (role,state)
 {if (findterminalp(state,game))
     {return findreward(role,state,game)};
  var actions = findlegals(role,state,game);
  var score = 0;
  for (var i=0; i<actions.length; i++)
     {var result = maxscore(role,simulate([actions[i]],state));
      if (result>score) {score = result}};
  return score}
```

Since all games in GGP competitions terminate, this procedure always halts; and it is easy to see that it produces the utility for the specified role in the specified state.

5.4   SEQUENTIAL PLANNING

Compulsive deliberation is wasteful in that computations are repeated unnecessarily. Once a player is able to find a path to a terminal state with maximal reward, it should not have to repeat that computation on every step. Sequential planning is the antithesis of compulsive deliberation in which no work is repeated. Once a sequential planner finds a good path, it simply saves the sequence of actions along that path and then executes the actions step by step until the game is done without any further deliberation.

A *sequential plan* for a single player game is any sequence of feasible actions. A sequential plan is *legal* if and only if every action in the sequence is legal in the state in which that action is performed. A plan is *complete* if and only if it leads from the initial state of the game to a terminal state. A plan is *minimal* if and only if none of the intermediate states is terminal.

Consider the 8-puzzle game described above. The sequential plans shown below are all legal, complete, and minimal. The actions in the sequences shown are all legal. Each leads from the initial state to a terminal state. And none of the intermediate states produced during the execution is terminal.

```
[right,down,right,down]
[right,down,left,right,right,down]
[right,down,left,right,left,right,right,down]
```

Note that this definition does not require that the sequence of actions produces the best possible result. This allows us to compare good plans and bad plans. We say that one sequential plan is *better than* another if and only if the reward associated with the first plan is greater than the reward associated with the second plan; and we define a sequential plan to be *optimal* if and only if there is no better sequential plan.

The table below shows that rewards associated with the plans shown above. Clearly, the second plan is better than the first, and the last three plans are optimal.

```
[right,left,right,left,right,left,right,left]    40
[right,right,left,right,left,right,left,right]    50
[right,down,right,down]                          100
[right,down,left,right,right,down]               100
[right,down,left,right,left,right,right,down]    100
```

Sequential planning is the process of finding sequential plans. A sequential planner is said to be *admissible* if and only if it returns an optimal sequential plan.

A *sequential planning player* is one that produces an optimal sequential plan and then executes the steps of that plan during game play. The planning is usually done during the start-up period of the game, but it can also be done during regular game play. It is also possible to mix sequential planning with other techniques. For example, in the case of large games, a player might play randomly during the initial part of a game and then switch to sequential planning once the game tree becomes small enough. Of course, in this last case, the player's ability to succeed depends on the strategy used before sequential planning commences.

Definitions for the basic methods for pure sequential planning are shown below. The *start* method invokes a procedure, called bestplan, to produce a sequential plan and stores this plan for later use. On each *play* of the game, the player performs the corresponding action in the plan. The *stop* and *abort* methods are the same as for the other players we have seen thus far.

```
var plan;
var step;

function start (id,player,rules,start,play)
 {game = rules;
  role = player;
  roles = findroles(game);
  state = findinits(game);
```

```
        plan = bestplan(role,state)[1];
        step = 0;
        return `ready'}

   function play (id,move)
    {var action = plan[step];
     step = step + 1;
     return action}
```

A depth-first search is conceptually the simplest approach to sequential planning. The procedure takes a player and a state as arguments and returns the corresponding utility and plan as value. The procedure takes the form of a recursive exploration of the game tree. If the state supplied as argument is terminal, the output is just the player's reward for that state and the empty plan. Otherwise, the output is the maximum of the utilities of the states resulting from executing any of the player's legal actions in the given state and the corresponding plan.

```
   function bestplan (role,state)
    {if (terminalp(state)) {return seq(findreward(role,state,game),[])};
     var actions = findlegals(role,state,game);
     var result = bestplan(role,findnext(roles,[actions[0]],state,game));
     var score = result[0];
     var plan = result[1];
     plan[plan.length] = actions[0];
     for (var i=1; i<actions.length; i++)
      {var result = bestplan(role,findnext(roles,[actions[i]],state,game));
         if (result[0]>score}
             {score = result[0];
              plan = result[1];
              plan[plan.length] = actions[i]};
     return seq(score,plan)}
```

Note that this procedure may not produce the shortest plan. However, it is guaranteed to produce an optimal plan as defined above.

PROBLEMS

Problem 5.1: What is the reward associated with each of the 8 Puzzle states shown below?

(a)
	1	2
3	4	5
6	7	8

(b)
1	2	3
4		5
6	7	8

(c)
	1	3
4	2	5
7	8	6

(d)
1	2	3
4	5	6
7	8	

Problem 5.2: Consider the 8-puzzle with the initial state shown below, and assume that there are exactly two steps left in the game, i.e., the game ends after exactly two additional actions.

1	2	3
4	5	6
7		8

(a) What is the player's reward in this state?

(b) What is the utility of this state?

(c) How many nodes are in the game tree?

(d) How many distinct states are in the game tree?

(e) What is the maximum number of nodes examined by compulsive deliberation?

Problem 5.3: Consider the 8-puzzle introduced in the notes. The initial state of the game is shown below on the left, and the ideal arrangement is shown on the right.

	1	3
4	2	5
7	8	6

1	2	3
4	5	6
7	8	

Recall that there are 4 feasible actions—up, down, left, right—and any of these actions is legal so long as the empty tile remains within the 3×3 grid. The reward for a state is the sum of rewards for each tile, with 10 points being awarded for each numbered tile in its ideal position and 20 points being awarded if the empty cell is in its proper position.

(*a*) Which of the following sequential plans are legal?

```
[right,right,down,down]
[right,right,right,down]
[right,down,right,down,left,up,right,down]
[right,down,right,left,right,left,right,down]
```

(*b*) Which of the following sequential plans are complete?

```
[right,right,down,down]
[right,down,right,left,right,left,right,left]
[right,down,right,left,right,left,right,down]
[right,down,right,down,left,up,right,down]}
```

(*c*) Which of the following sequential plans are minimal?

```
[right,right,down,down]
[right,down,right,left,right,left,right,left]
[right,down,right,left,right,left,right,down]
[right,down,right,down,left,up,right,down]
```

(*d*) Which of the following sequential plans are optimal?

```
[right,down,right,down]
[right,down,right,left,right,left,right,left]
[right,down,right,left,right,left,right,down]
```

CHAPTER 6

Small Multiple-Player Games

6.1 INTRODUCTION

Having dealt with small single-player games, we turn now to small multiple-player games. In most cases, the other players are general game playing programs or humans. However, in some cases, the other players represent uncertainty in the game itself. For example, it is common to model some card games by representing a randomly shuffled deck of cards as an additional player in the game, one that deals or reveals cards as the game progresses.

Multiple-player games are more complicated than single-player games because the state resulting from a player's actions can depend on the actions of the other players. No player can directly control the actions of other players; and so, in making its choices, a player must consider all possible actions of the other players.

Before proceeding, it is worth emphasizing that games need not be *fixed sum*. In a fixed sum game, the total number of points is fixed. (When this number is zero, such games are usually said to be *zero-sum*.) In order for one player to get more points, some other player must lose points. For this reason, fixed sum games are necessarily competitive. In general, game playing, there is no such restriction. Some games are competitive; but others are cooperative—it may be that the only way for one player to get a higher reward is to help the other players get higher rewards as well.

While it is possible, in some multiple-player games, to find sequential plans that produce maximal rewards, this is rarely the case. In order to achieve an optimal reward, it is frequently necessary for a player to conditionalize its actions on the state of the game. This is a situation where compulsive deliberation works well.

In this chapter, as in the preceding chapter, we look at settings in which there is sufficient time for players to search the game tree entirely. That said, as in single-player games, it is sometimes possible to find optimal actions even without searching the entire game tree.

We begin this chapter with a procedure called *Minimax*, and we then consider a more efficient variation called *Bounded Minimax*. We then turn to an even more efficient procedure called *Alpha-Beta Search*, which produces the same results but eliminates some of the needless computation of minimax.

6.2 MINIMAX

In general game playing, a player may choose to make assumptions about the actions of the other players. For example, a player might want to assume that the other players are behaving rationally.

By eliminating irrational actions on the part of the other players, a player can decrease the number of possibilities it needs to consider.

Unfortunately, in general game playing, as currently constituted, no player knows the identity of the other players. The other players might be irrational or they might behave the same as the player itself. Since there is no information about the other players, many general game players take a pessimistic approach—they assume that the other players will perform the worst possible actions. This pessimistic approach is the basis for a game-playing technique called *Minimax*.

In the case of a one-step game, Minimax chooses an action such that the value of the resulting state for *any* opponent action is greater than or equal to the value of the resulting state for any other action. In the case of a multiple-step game, Minimax goes to the end of the game and *backs up* values.

We can think about Minimax as search of a bipartite tree consisting of alternating *max nodes* and *min nodes*. See the example in Figure 6.1. The max nodes (shown in beige) represent the choices of the player while the min nodes (shown in grey) represent the choices of the other players.

Note that, although we have separated the choices of the player and its opponents, this does not mean that play alternates between the opponents or that the opponents know the player's action. The player and its opponents make their choices simultaneously, without knowledge of each other's choices.

The value of a max node for a player is either the utility of that state if it is terminal or the maximum of all values for the min nodes that result from its legal actions. The value of a min node is the minimum value that results from any legal opponent action.

The following game tree in Figure 6.2 illustrates this. The nodes at the bottom of the tree are terminal states, and the values are the player's goal values for those states. The values shown in the other nodes are computed according to the rules just stated. For example, the value of the minnode at the lower left is 1 because that is the minimum of the values of its maxnodes below it, viz. 1 and 2. The value of the minnode next to that minnode is 3 because that is the value of the two maxnodes below it, viz. 3 and 4. The value of the maxnode above these two minnodes is 3 because that is the maximum of the values of the two minnodes. And so forth.

The following procedure is a simple implementation of a player that uses minimax to evaluate states. The implementation is similar to that of the compulsive deliberation player introduced in the preceding chapter. The `start` handler merely records relevant information in the player's global variables. The `play` handler simulates the previous move to obtain the current state and then finds the best action for it to perform in that state. In this case, it uses the `bestmove` subroutine to obtain this action.

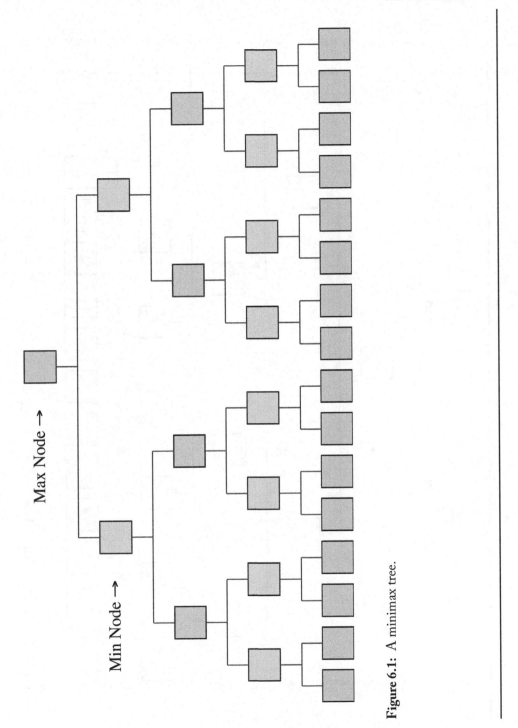

Figure 6.1: A minimax tree.

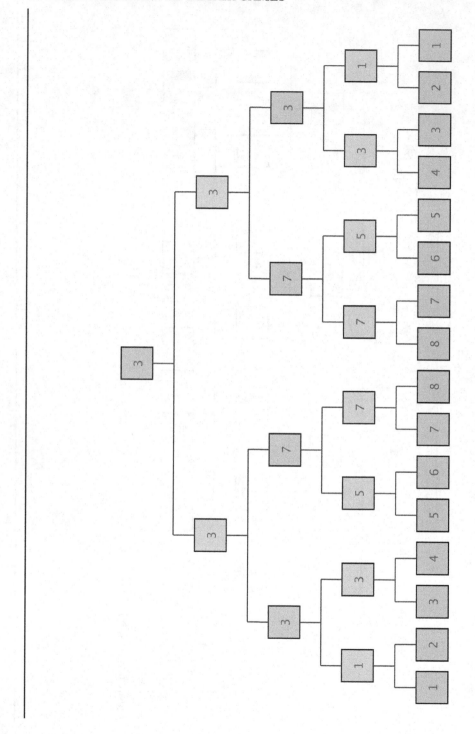

Figure 6.2: A minimax tree with values.

```
function start (id,player,rules,sc,pc)
 {game = rules;
  role = player;
  roles = findroles(game)
  state = findinits(game);
  startclock = sc;
  playclock = pc;
  return `ready'}

function play (id,move)
 {state = simulate(move,state);
  return bestmove(role,state)}
```

The main difference between the bestmove subroutine for single-player games and the bestmove for multiple-player games is the way scores are computed. Rather than comparing subsequent states, it compares min nodes as described above.

```
function bestmove (role,state)
 {var actions = findlegals(role,state,game);
  var action = actions[0];
  var score = 0;
  for (var i=0; i<actions.length; i++)
      {var result = minscore(role,actions[i],state);
       if (result>score) {score = result; action = actions[i]}};
  return action}
```

The minscore subroutine takes a role, an action of that role, and a state as arguments and produces the minimum values for the given role associated with the given action for any of the opponent's legal actions in the given state. (The findopponent subroutine here uses the game description to compute the other player in a two-player game.)

```
function minscore (role,action,state)
 {var opponent = findopponent(role,game);
  var actions = findlegals(opponent,state,game);
  var score = 100;
  for (var i=0; i<actions.length; i++)
      {var move;
       if (role==roles[0]) {move = [action,actions[i]]}
           else {move = [actions[i],action]}
       var newstate = findnext(move,state,game);
```

```
            var result = maxscore(role,newstate);
            if (result<score)  {score = result}};
      return score}
```

The `maxscore` subroutine, which is called by `minscore`, takes a role and a state as arguments. It conducts a recursive exploration of the game tree below the given state. If the state is terminal, the output is just the role's reward for that state. Otherwise, the output is the maximum of the utilities of the min nodes associated with the player's legal actions in the given state.

```
function maxscore (role,state)
  {if (findterminalp(state,game)) {return findreward(role,state,game)};
   var actions = findlegals(role,state,game);
   var score = 0;
   for (var i=0; i<actions.length; i++)
      {var result = minscore(role,actions[i],state);
       if (result>score) {score = result}};
   return score}
```

6.3 BOUNDED MINIMAX SEARCH

One disadvantage of the Minimax procedure described in the preceding section is that it examines the entire game tree in all cases. While this is sometimes necessary, there are cases where it is possible to get the same result without examining the entire game tree. For example, if in processing a state the `maxscore` subroutine finds an action that produces 100 points, it does not need to look at any additional actions since it cannot do better; and if the `minscore` subroutine finds an action that produces 0 points, it does not need to look at any additional actions since it cannot get the score any lower.

Bounded Minimax is just the Minimax procedure just discussed. Rather than processing all actions on every node, it checks first for these bounds; and, if they occur on any node, it terminates its examination and returns the corresponding value.

As an example of this, consider the game tree shown in Figure 6.3. The nodes with values are those examined by Bounded Minimax. The other nodes are not examined at all and do not need to be examined.

It is easy to adapt the basic Minimax code to do Bounded Minimax. All we need to do is to put conditionals in the inner loops of bestmove and `maxscore` and `minscore`, as shown below.

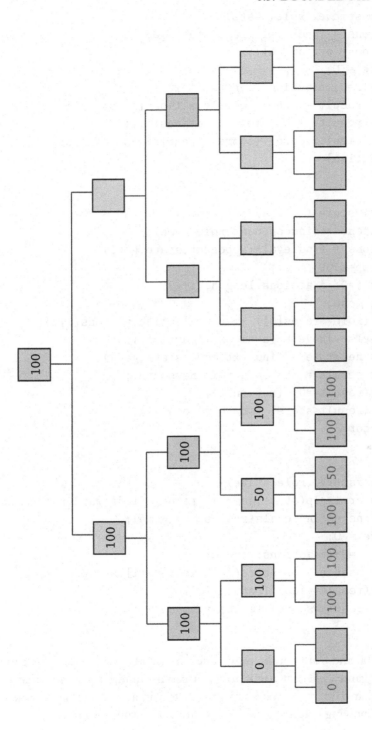

Figure 6.3: A minimax tree with values computed by Bounded Minimax.

```
function bestmove (role,state)
 {var actions = findlegals(role,state,game);
  var action = actions[0];
  var score = 0;
  for (var i=0; i<actions.length; i++)
      {var result = minscore(role,actions[i],state);
       if (result==100) {return actions[i]};
       if (result>score) {score = result; action = actions[i]}};
  return action}


function minscore (role,action,state)
 {var opponent = findopponent(role,game);
  var actions = findlegals(opponent,state,game);
  var score = 100;
  for (var i=0; i<actions.length; i++)
      {var move;
       if (role==roles[0]) {move = [action,actions[i]]}
          else {move = [actions[i],action]}
       var newstate = findnext(move,state,game);
       var result = maxscore(role,newstate);
       if (result==0) {return 0};
       if (result<score) {score = result}};
  return score}


function maxscore (role,state)
 {if (findterminalp(state,game)) {return findreward(role,state,game)};
  var actions = findlegals(role,state,game);
  var score = 0;
  for (var i=0; i<actions.length; i++)
      {var result = minscore(role,actions[i],state);
       if (result==100) {return 100};
       if (result>score) {score = result}};
  return score}
```

Note that 100 and 0 are not the only values that can be used here. For example, if a player is in a satisficing game, where it needs to get a certain minimum score, then it can use that threshold rather than 100. If a player simply wants to win a fixed sum game, then it can use 51 as the threshold, knowing that if it gets this amount it has won the game.

6.4   ALPHA-BETA SEARCH

While Bounded Minimax helps avoid some wasted work, we can do even better. Consider the game tree on the next page.

In this case, unlike the examples seen earlier, there are many terminal values that are not 0 or 100. In determining its maximum score for the top node of this tree, a Minimax player, even a Bounded Minimax player, would examine the entire tree. However, not all of this work is necessary.

Alpha-Beta Search is a variation on Bounded Minimax that eliminates such wasted work by computing bounds dynamically and passing them along as parameters. One bound, called alpha, is the best score the player has seen thus far. The other bound, called beta, is the worst score the player has seen. In examining new nodes, alpha-beta search uses these bounds to decide whether to look at further nodes.

If the partial result at a min node is less than alpha, then there is no point in examining other descendants of that node since it could only decrease this value and the player would not take that choice given that it has a higher value elsewhere.

Analogously, if the partial result at a max node is greater than beta, then there is no point in considering other options since they can only increase the score and the player's opponents would not allow that since they know they can keep the value to no more than beta.

The following is an implementation of `maxscore` and `minscore` for an alpha-beta player.

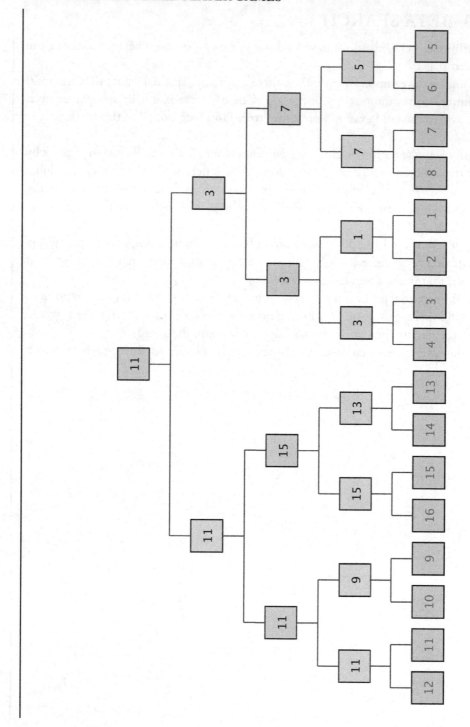

Figure 6.4: A sample minimax tree.

```
function maxscore (role, state, alpha, beta)
  {if (findterminalp(state,game)) {return findreward(role,state,game)};
   var actions = findlegals(role,state,game);
   for (var i=0; i<actions.length; i++)
       {var result = minscore(role,actions[i],state,alpha,beta);
        alpha = max(alpha,result);
        if (alpha>=beta) then {return beta}};
   return alpha}

function minscore (role, action, state,alpha,beta)
  {var opponent = findopponent(role,game);
   var actions = findlegals(opponent,state,game);
   for (var i=0; i<actions.length; i++)
       {var move;
        if (role==roles[0]) {move = [action,actions[i]]}
           else {move =[actions[i],action]}
        var newstate = findnext(move,state,game);
        var result = maxscore(role,newstate,alpha,beta);
        beta = min(beta, result);
        if (beta<=alpha) {return alpha}};
   return beta}
```

Now let's apply the maxscore procedure to the tree shown above with initial value 0 and 100 for alpha and beta. In the tree below, we have written in values produced by the alpha-beta procedure in this case, and we have left the other nodes blank (next page).

In this particular case, the improvement of Alpha-Beta over Minimax is modest. However, in general, Alpha-Beta Search can save a significant amount of work over full Minimax. In the best case, given a tree with branching factor b and depth d, Alpha-Beta Search needs to examine at most $O(b^{d/2})$ nodes to find the maximum score instead of $O(b^d)$. This means that an Alpha-Beta player can look ahead twice as far as a Minimax player in the same amount of time. Looked at another way, the effective branching factor of a game in this case is $\sqrt{b}$ instead of b. It would be the equivalent of searching a tree with just 5 moves instead of 25 moves.

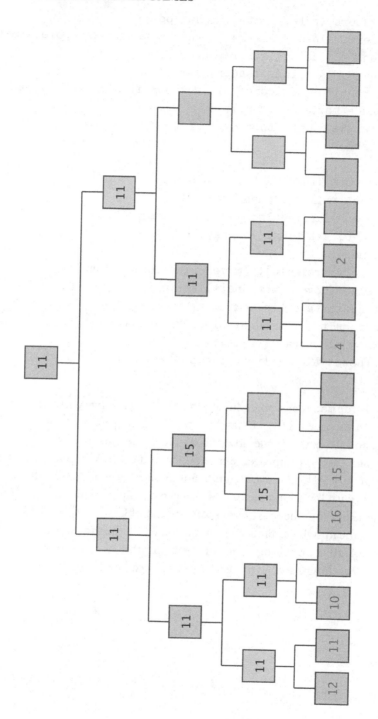

Figure 6.5: A minimax tree with values computed by Alpha–Beta Search.

PROBLEMS

Problem 6.1: Fill in the minimax values for the non-terminal nodes in the game tree in Figure 6.6. Max nodes are beige; min nodes are grey.

Problem 6.2: Assign the utility values 1, 2, ..., 16 to the 16 terminal nodes in the game tree in Figure 6.7 so that (1) no two terminal nodes have the same value and (2) the minimax value of the top node is 8.

Problem 6.3: Fill in the alpha-beta values for the non-terminal nodes in the game tree in Figure 6.8. Put an X in any node that alpha-beta does not examine.

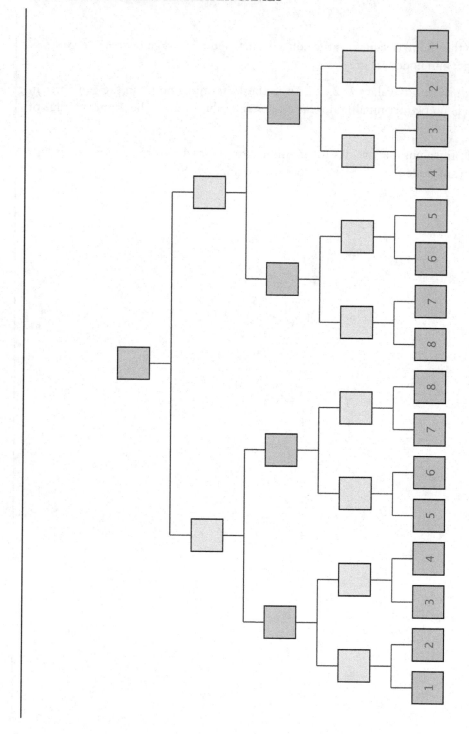

Figure 6.6: Problem 6.1.

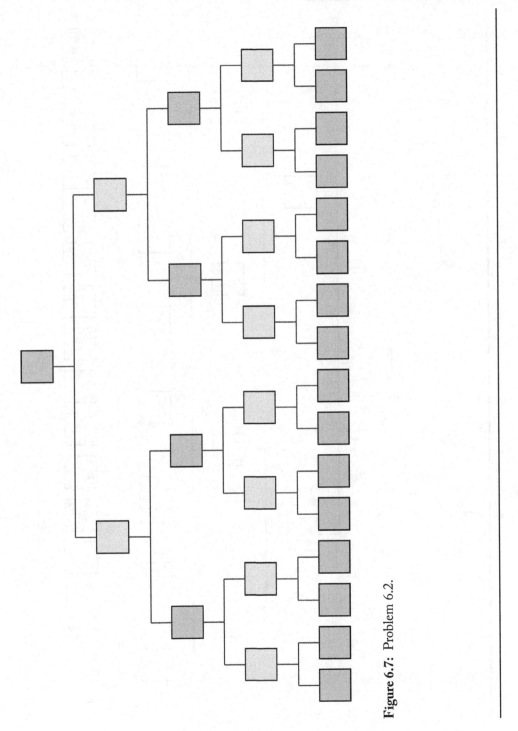

Figure 6.7: Problem 6.2.

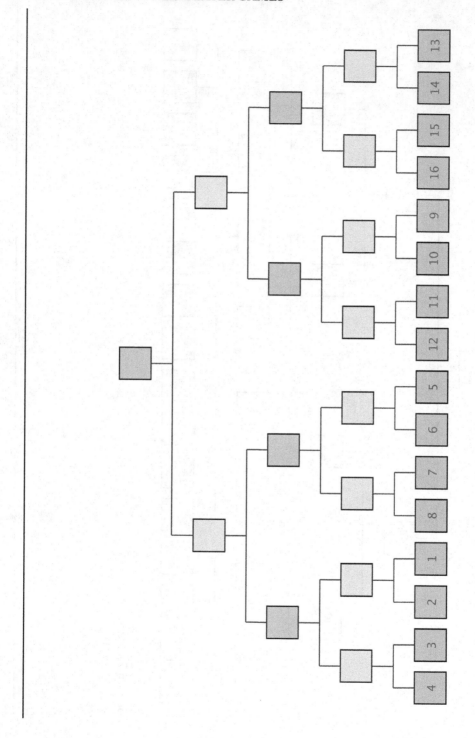

Figure 6.8: Problem 6.3.

CHAPTER 7

Heuristic Search

7.1 INTRODUCTION

In the last two chapters, we looked at approaches to playing small games, i.e., games for which there is sufficient time for a complete search of the game tree. Unfortunately, most games are not so small, and complete search is usually impractical. In this chapter, we look at a variety of techniques for incomplete search. We begin with Depth-Limited Search, then Fixed-Depth Heuristic Search, and finally Variable-Depth Heuristic Search. In the next chapter, we examine probabilistic methods for dealing with incomplete search.

7.2 DEPTH-LIMITED SEARCH

The simplest way of dealing with games for which there is insufficient time to search the entire game tree is to limit the search in some way. In Depth-Limited Search, the player explores the game tree to a given depth. A legal player is a special case of Depth-Limited Search where the depth is effectively zero.

The implementation of Depth-Limited Search is a simple variation of the implementation of the minimax player described in the preceding chapter. See below. One difference is the addition of a level parameter to maxscore and minscore. This parameter is incremented on each recursive call in minscore. If the player reaches the depth-limit at a non-terminal state, it simply returns 0, a conservative lower bound on its utility.

```
function maxscore (role,state,level)
  {if (findterminalp(state,game)) {return findreward(role,state,game)};
   if (level>=limit) {return 0};
   var actions = findlegals(role,state,game);
   var score = 0;
   for (var i=0; i<actions.length; i++)
       {var result = minscore(role,actions[i],state,level);
        if (result==100) {return 100};
        if (result>score) {score = result}};
   return score}
```

```
function minscore (role,action,state,level)
 {var opponent = findopponent(role,game);
  var actions = findlegals(opponent,state,game);
  var score = 100;
  for (var i=0; i<actions.length; i++)
      {var move;
        if (role==roles[0]) {move = [action,actions[i]]}
           else {move = [actions[i],action]}
        var newstate = findnext(move,state,game);
        var result = maxscore(role,newstate,level+1);
        if (result==0) {return 0};
        if (result<score) {score = result}};
   return score}
```

The most obvious problem with Depth-Limited Search is that the conservative estimate of utility for non-terminal states is not very informative. In the worst case, none of the states at a given depth may be terminal, in which case the search provides no discriminatory value. We discuss some ways of dealing with this problem in the next sections and the next chapter.

Another problem with Depth-Limited Search is that a player may not be able to determine a suitable depth-limit in advance. Too low and the player will not search as much as it could. Too high and the search may not terminate in time.

One solution to this problem is to use breadth-first search rather than depth-first search. The downside of this is the amount of space that this consumes on very large trees, in many cases exceeding the storage limits of the computer.

An alternative solution to this problem is to use an iterative deepening approach to game tree exploration, exploring the game tree repeatedly at increasing depths until time runs out. As usual with iterative deepening, this is wasteful in that portions of the tree may be explored multiple times. However, as usual with iterative deepening, this waste is usually bounded by a small constant factor.

7.3   FIXED-DEPTH HEURISTIC SEARCH

One way of dealing with the conservative nature of Depth-Limited Search is to improve upon the arbitrary 0 value returned for nonterminal states. In fixed-depth heuristic search, this is accomplished by applying a heuristic evaluation function to non-terminal states. Such functions are based on features of states, and so they can be computed without examining entire game tree.

Examples of such heuristic functions abound. For example, in Chess, we often use piece count to compare states, with the idea that, in the absence of immediate threats, having more material is better than having less material. Similarly, we sometimes use board control, with the idea that having control of the center of the board is more valuable than controlling the edges or corners.

The downside of using heuristic functions is that they are not necessarily guaranteed to be successful. They may work in many cases but they can occasionally fail, as happens, for example in Chess, when a player is checkmated even though it has more material and a better board control. Still, games often admit heuristics that are useful in the sense that they work more often than not.

While, for specific games, such as Chess, programmers are able to build in evaluation functions in advance, this is unfortunately not possible for general game playing, since the structure of the game is not known in advance. Rather, the game player must analyze the game itself in order to find a useful evaluation function. In a later chapter, we discuss how to find such heuristics.

That said, there are some heuristics for game playing that have arguable merit across all games. In this section, we examine some of these heuristics. We also show how to build game players that utilize these general heuristics.

The implementation of a general game player based on fixed-depth heuristic search is a simple variation of the fixed-depth search player just described. See below. The difference comes in the maxscore procedure. Rather than returning 0 on non-terminal states, the procedure returns the value of a subroutine evalfn, which gives a heuristic value for the state.

```
function maxscore (role,state,level)
 {if (findterminalp(state,game)) {return findreward(role,state,game)};
  if (level>=limit) {return evalfn(role,state)};
  var actions = findlegals(role,state,game);
  var score = 0;
  for (var i=0; i<actions.length; i++)
     {var result = minscore(role,actions[i],state,level);
      if (result==100) {return 100};
      if (result>score) {score = result}};
  return score}
```

There are various ways of defining evalfn. In the following paragraphs, we look at just a few of these—mobility, focus, and goal proximity.

Mobility is a measure of the number of things a player can do. This could be the number of actions available in the given state or n steps away from the given state. Or it could be the number of states reachable within n steps. (This could be different from the number of actions since multiple action sequences could lead to the same state. All roads lead to Rome.)

A simple implementation of the mobility heuristic is shown below. The method simply computes the number of actions that are legal in the given state and returns as value the percentage of all feasible actions represented by this set of legal actions.

```
function mobility (role,state)
 {var actions = findlegals(role,state,game);
  var feasibles = findactions(role,game);
  return (actions.length/feasibles.length * 100)}
```

Focus is a measure of the narrowness of the search space. It is the inverse of mobility. Sometimes it is good to focus to cut down on search space. Often better to restrict opponents' moves while keeping one's own options open.

```
function focus (role,state)
 {var actions = findlegals(role,state,game);
  var feasibles = findactions(role,game);
  return (100 - actions.length/feasibles.length * 100)}
```

Goal proximity is a measure of how similar a given state is to desirable terminal state. There are various ways this can be computed.

One common method is to count how many propositions that are true in the current state are also true in a terminal state with adequate utility. The difficulty of implementing this method is obtaining a set of desirable terminal states with which the current state can be compared.

Another alternative is to use the utility of the given state as a measure of progress toward the goal, with the idea being that the higher utility, the closer the goal. Of course, this is not always true. However, in many games the goal values are indeed *monotonoic*, meaning that values do increase with proximity to the goal. Moreover, it is sometimes possible to compute this by a simple examination of the game description, using methods we describe in later chapters.

None of these heuristics is guaranteed to work in all games, but all have strengths in some games. To deal with this fact, some designers of GGP players have opted to use a weighted combination of heuristics in place of a single heuristic. See the formula below. Here each f_i is a heuristic function, and w_i is the corresponding weight.

$$f(s) = w_1 \times f_1(s) + \ldots + w_n \times f_n(s)$$

Of course, there is no way of knowing in advance what the weights should be, but sometimes playing a few instances of a game (e.g., during the start clock) can suggest weights for the various heuristics.

7.4 VARIABLE DEPTH HEURISTIC SEARCH

As we discussed in the preceding section, heuristic search is not guaranteed to succeed in all cases. Failing is never good. However, it is particularly embarrassing in situations where just a little more search would have revealed significant changes in the player's circumstances, for better or worse. In the research literature, this is often called a *horizon problem*.

As an example of a horizon problem in Chess, consider a situation where the players are exchanging piece, with white capturing black's pieces and vice versa. Now imagine cutting off the search at an arbitrary depth, say 2 captures each. At this point, white might believe it has an advantage since it has more material. However, if the very next move by black is a capture of the white queen, this evaluation could be misleading.

A common solution to this problem is to forego the fixed depth limit in favor of one that is itself dependent on the current state of affairs, searching deeper in some areas of the tree and searching less deep in other areas.

In Chess, a good example of this is to look for quiescence, i.e., a state in which there are no immediate captures.

The following is an implementation of a variable-depth heuristic. This version of maxscore differs from the fixed-depth version in that there is a subroutine (here called expfn) that is called to determine whether the current state and/or depth meets an appropriate condition. If so, the tree expansion terminates; otherwise, the player expands the state.

```
function maxscore (role,state,level)
  {if (findterminalp(state,game)) {return findreward(role,state,game)};
   if (!expfn(role,state,level)) {return evalfn(role,state)};
   var actions = findlegals(role,state,game);
   var score = 0;
   for (var i=0; i<actions.length; i++)
       {var result = minscore(role,actions[i],state,level);
        if (result==100) {return 100};
        if (result>score) {score = result}};
   return score}
```

The challenge in variable-depth heuristic search is finding an appropriate definition for expfn. One common technique is to focus on differentials of heuristic functions. For example, a significant *change* in mobility or goal proximity might indicate that further search is warranted whereas actions that do not lead to dramatic changes might be less important.

PROBLEMS

Problem 7.1: Consider the single-player game tree shown on the next page, and answer the questions that follow.

(*a*) What is the minimax value of the tree?

(*b*) What is the value returned by Depth-Limited Search with a depth limit of 3?

(*c*) How many nodes are examined by depth-first search with a depth-limit of 3, i.e., how many times is maxscore called?

(*d*) How many nodes are examined by breadth-first search with a depth-limit of 3, i.e., how many times is `maxscore` called?

(*e*) How many nodes are examined by iterative-deepening search with a depth-limit of 3 and a depth increment of 1, i.e., how many times is `maxscore` called?

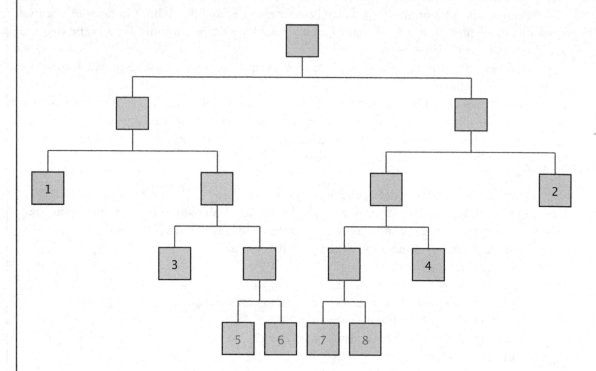

Problem 7.2: Consider the two-player game tree shown below. The values on the max nodes are actual goal values for the associated states as given in the game description, *not* state utilities determined by game tree search.

(*a*) What is the state utility of the top of the tree (as determined, for example, by Minimax)?

(*b*) Now consider a player using fixed-depth heuristic search with depth limit 1. How many max nodes are searched in evaluating the top node in this tree (i.e., how many times is `maxscore` called)?

(*c*) Suppose the player uses a goal proximity heuristic with state reward as the heuristic value for non-terminal states. What is the minimum final reward for this player in this game?

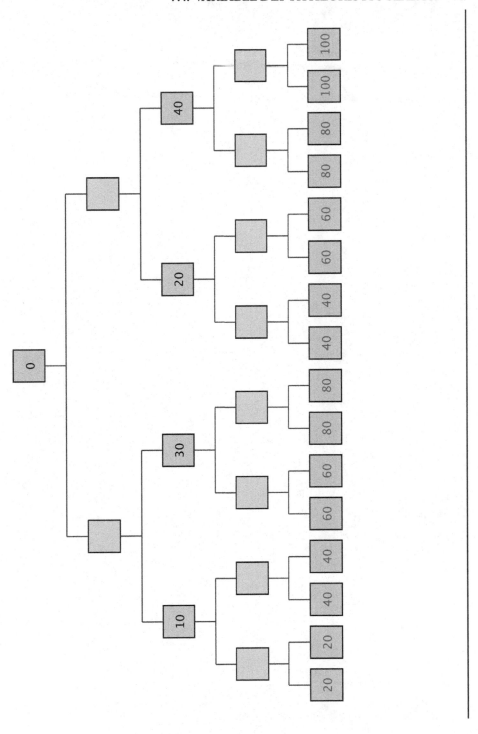

CHAPTER 8

Probabilistic Search

8.1 INTRODUCTION

In the preceding chapter, we examined various approaches to incomplete search of game trees. In each approach, the evaluation of states is based on local properties of those states (i.e., properties that do not depend on the game tree as a whole). In many games, there is no correlation between these local properties and the likelihood of success in completing a game successfully.

In this chapter, we look at some alternative methods based on probabilistic analysis of game trees. In the next section, we examine an approach based on Monte Carlo game simulation. In the subsequent section, we look at a more sophisticated variation called Monte Carlo Tree Search.

8.2 MONTE CARLO SEARCH

The basic idea of Monte Carlo Search (MCS) is simple. In order to estimate the value of a non-terminal state, we make some *probes* from that state to the end of the game by selecting random moves for the players. We sum up the total reward for all such probes and divide by the number of probes to obtain an *estimated utility* for that state. We can then use these expected utilities in comparing states and selecting actions.

Monte Carlo can be used in compulsive deliberation fashion to evaluate the immediate successors of the current state. However, it can also be used as an evaluation function for heuristic search. The former approach can then be seen as a special case of the latter where the depth limit for expansion is set to 0. In the latter case, search takes the form of two phases of search: the expansion phase and the probe phase.

The expansion phase of Monte Carlo is the same as bounded depth heuristic search. The tree is explored until some fixed depth is reached. The tree shown below illustrates this process.

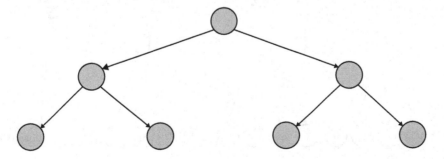

The probe phase of Monte Carlo takes the form of exploration from each of the fringe states reached in the expansion phase, for each making random probes from there to a terminal state. See below.

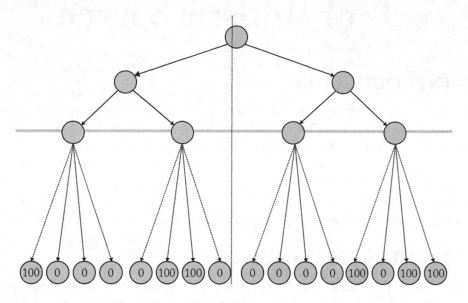

The values produced by each probe are added up and divided by the number of probes for each state to obtain an expected utility for that state. These expected utilities are then compared to determine the relative utilities of the fringe states produced at the end of the expansion phase.

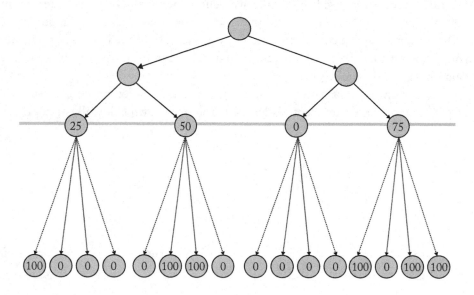

The following procedure is a simple implementation of the two-phase Monte Carlo method, with 4 probes per state. The implementation is similar to that of the heuristic search player introduced in the preceding chapter. The main difference is that the Monte Carlo method is used to evaluate states rather than general heuristics.

```
function maxscore (role,state,level)
 {if (findterminalp(state,game)) {return findreward(role,state,game)};
  if (level>levels) {return montecarlo(state)};
  var actions = findlegals(role,state,game);
  var move = seq();
  var score = 0;
  for (var i=0; i<actions.length; i++)
      {var result = minscore(role,actions[i],state,level);
       if (result==100) {return 100};
       if (result>score) {score = result}};
  return score}


function maxscore (role,state,level)
 {if (findterminalp(state,game)) {return findreward(role,state,game)};
  if (level>levels) {return montecarlo(role,state,4)};
  var actions = findlegals(role,state,game);
  var score = 0;
  for (var i=0; i<actions.length; i++)
      {var result = minscore(role,actions[i],state,level);
       if (result==100) {return 100};
       if (result<score) {score = result}};
  return score}


function montecarlo (role,state,count)
 {var total = 0;
  for (var i=0; i<count; i++)
      {total = total + depthcharge(role,state)};
  return total/count}


function depthchargescore (role,state)
 {var total = 0;
  for (var i=0; i<count; i++)
      {total = total + depthcharge(role,state)};
  return total/count}
```

```
function depthcharge (role,state)
 {if (findterminalp(state,game)) {return findreward(role,state,game)};
  var move = seq();
  for (var i=0; i<roles.length; i++)
      {var options = findlegals(roles[i],state,game);
       move[i] = randomelement(options)};
  var newstate = simulate(move,state);
  return depthcharge(role,newstate)}
```

Note that, in the probe phase of Monte Carlo, only one action is considered for each player on each step. Also, there is no additional processing (e.g., checking whether states have been previously expanded). Consequently, making probes in Monte Carlo is fast, and this enables players to make many such probes.

One downside on the Monte Carlo method is that it can be optimistic. It assumes the other players are playing randomly when in fact it is possible that they know exactly what they are doing. It does not help if most of the probes from a position in Chess lead to success if one leads to a state in which one's player is checkmated and the other player sees this. This issue is addressed to some extent in the MCTS method described below.

Another drawback of the Monte Carlo method is that it does not take into account the structure of a game. For example, it may not recognize symmetries or independencies that could substantially decrease the cost of search. For that matter, it does not even recognize boards or pieces or piece count or any other features that might form the basis of game-specific heuristics. These issues are discussed further in the chapters to follow.

Even with those drawbacks, the Monte Carlo method is quite powerful. Prior to its use, general game players were at best interesting novelties. Once players started using Monte Carlo, the improvement in game play was dramatic. Suddenly, automated general game players began to perform at a high level. Using a variation of this technique, CadiaPlayer won the International General Game Playing competition three times. Almost every general game playing program today includes some version of Monte Carlo.

8.3 MONTE CARLO TREE SEARCH

Monte Carlo Tree Search (MCTS) is a variation of Monte Carlo Search. Both methods build up a game tree incrementally and both rely on random simulation of games; but they differ on the way the tree is expanded. MCS uniformly expands the partial game tree during its expansion phase and then simulates games starting at states on the fringe of the expanded tree. MCTS uses a more sophisticated approach in which the processes of expansion and simulation are interleaved.

MCTS processes the game tree in cycles of four steps each. After each cycle is complete, it repeats these steps so long as there is time remaining, at which point it selects an action based on the statistics it has accumulated to that point.

Selection. In the selection step, the player traverses the tree produced thus far to select an unexpanded node of the tree, making choices based on visit counts and utilities stored on nodes in the tree.

Expansion. The successors of the state chosen during the selection phase are added to the tree.

Simulation. The player simulates the game starting at the node chosen during the selection phase. In so doing, it chooses actions at random until a terminal state is encountered.

Backpropagation. Finally, the value of the terminal state is propagated back along the path to the root node and the visit counts and utilities are updated accordingly.

An implementation of the MCTS selection procedure is shown below. If the initial state has not been seen (i.e., it has 0 visits), then it is selected. Otherwise, the procedure searches the successors of the node. If any have not been seen, then one of the unseen nodes is selected. If all of the successors have been seen before, then the procedure uses the `selectfn` subroutine (described below) to find values for those nodes and chooses the one that maximizes this value.

```
function select (node)
 {if (node.visits==0) {return node};
  for (var i=0; i<node.children.length; i++)
      {if (node.children[i].visits==0) {return node.children[i]}};
  score = 0;
  result = node;
  for (var i=0; i<node.children.length; i++)
      {var newscore = selectfn(node.children[i]);
       if (newscore>score)
          {score = newscore; result=node.children[i]}};
  return select(result)}
```

One of the most common ways of implementing `selectfn` is UCT (Upper Confidence bounds applied to Trees). A typical UCT formula is $vi + c^*sqrt(log\ np\ /\ ni)$. vi here is the average reward for that state. c is an arbitrary constant. np is the total number of times the state's parent was picked. ni is the number of times this particular state was picked.

```
function selectfn(node)
 {return node.utility+Math.sqrt(2*Math.log(node.visits)/
                                node.parent.visits)}
```

Of course, there are other ways that one can evaluate states. The formula here is based on a combination of exploitation and exploration. Exploitation here means the use of results on previously explored states (the first term). Exploration means expansion of as-yet unexplored states (the second term).

Expansion in MCTS is basically the same as that for MCS. An implementation for a single-player game is shown below.

```
function expand (node)
 {var actions = findlegals(role,node.state,game);
  for (var i=0; i<actions.length; i++)
      {var newstate = simulate(seq(actions[i]),state);
       var newnode = makenode(newstate,0,0,node,seq());
       node.children[node.children.length]=newnode};
  return true}
```

On large games with large time bounds, it is possible that the space consumed in this process could exceed the memory available to a player. In such cases, it is common to use a variation of the selection procedure in which no additional states are added to the tree. Instead, the player continues doing simulations and updating its numbers for already-known states.

Simulation for MCTS is essentially the same as simulation for MCS. So the same procedure can be used for both methods.

Backpropagation is easy. At the selected node, the method records a visit count and a utility. The visit count in this case is 1 since it was a newly processed state. The utility is the result of the simulation. The procedure then propagates to ancestors of this node. In the case of a single player game, the procedure adds 1 to the visit count of each ancestor and augments its total utility by the utility obtained on the latest simulation. See below. In the case of a multiple-player game the propagated value is the minimum of the values for all opponent actions.

```
function backpropagate (node,score)
 {node.visits = node.visits+1;
  node.utility = node.utility+score;
  if (node.parent) {backpropagate(node.parent,score)};
  return true}
```

Expanding the method to multiple-player games is tedious but poses no significant problems. Max nodes and min nodes must be differentiated. Expansion must create a bipartite tree, alternating between max nodes and min nodes. And backpropagation must be adjusted accordingly.

CHAPTER 9

Propositional Nets

9.1 INTRODUCTION

In the Introduction, we saw that it is possible to think of the dynamics of a game as a state graph. See, for example, the state graph in Fig. 9.1. A game is characterized by a finite number of states, a finite number of players, each with a finite number of actions. At each point in time the game is in one of the possible states; players choose from their possible actions; and, as the players perform their chosen actions, the game changes from one state to another.

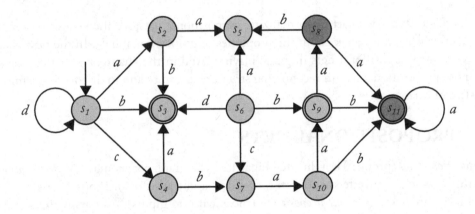

Figure 9.1: State Machine for a simple game.

However, we rarely think of states as monolithic entities. In practice, we typically characterize states in terms of propositions that are true in those states. As actions are performed, some propositions become true and others become false. This suggests a conceptualization of games as propositional nets rather than state machines. A propositional net is a graph in which propositions and actions are nodes rather than states and where these nodes are interleaved with nodes representing logical connectives and transitions, as suggested by the example shown in Figure 9.2.

One of the benefits of formalizing games as propositional nets is compactness. A set of n propositions corresponds to a set of 2^n states (all different combinations of truth values for the n propositions). Thus, it is often possible to characterize the dynamics of games with graphs that are much smaller than the corresponding state machines. For example, the propnet in Fig. 9.2, with just three propositions, corresponds to a state machine with eight states.

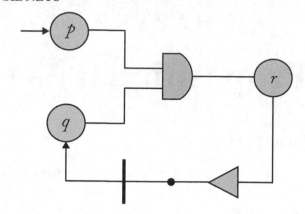

Figure 9.2: Propositional Net for a simple game.

In this chapter, we start by formalizing propositional nets; we then show how to describe games in this way and we talk about the properties of propositional nets. In the next chapter, we see how to use propositional nets in game playing. And in the chapters after that, we see how we can use propositional nets in recognizing structure in games and in discovering game-playing heuristics of various sorts.

9.2   PROPOSITIONAL NETS

A *propositional net* (propnet) is a directed bipartite hypergraph consisting of *propositions* alternating with *connectives* (inverters, and-gates, or-gates, and transitions). Propositions can be partitioned into three classes: *input propositions* (those with no inputs); *base propositions* (those with incoming arcs from transitions); and *view propositions* (those with incoming arcs from connectives other than transitions).

The propnet in Fig. 9.3 is an example. In this case, there are six propositions (the nodes labeled *a*, *b*, *p*, *q*, *r*, and *s*); and there are four connectives (the and-gate on the upper left, the inverter on the upper right, the or-gate on the lower right, and the transition on the lower left). Nodes *a* and *b* are input propositions; node *s* is a base proposition; and nodes *p*, *q*, and *r* are view propositions.

An *input marking* is a function from the input propositions of a propositional net to boolean values. A *base marking* is a function from the base propositions of a propositional net to boolean values. A *view marking* is a function from the view propositions of a propositional net to boolean values.

Given a propnet, an input marking and a base marking determine a unique view marking for that propnet. This is based on the types of gates leading into the view propositions. The output of an inverter is true if and only if its input is false. The output of an and-gate is true if and only

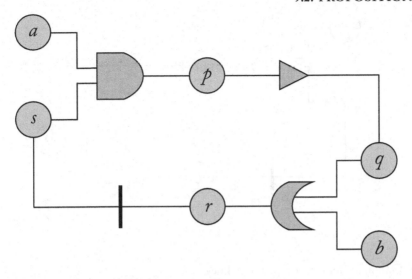

Figure 9.3: Sample propositional net.

all of its inputs are true. The output of an or-gate is true if an only if at least one of its inputs is true.

As an example, consider the propnet in Fig. 9.3. Suppose we had an input marking that assigned *a* the value true and *b* the value false, and suppose we had a base marking that assigned *s* the value true. Then, the view marking for *p* would be true; the view marking for *q* would be false; and the view marking for *r* would be false. At this point, we have values for all of the view propositions in the propnet. See Fig. 9.4. Here we have written 1 for true and 0 for false.

Transitions are the basis for dynamics in a propnet. Consider a step in the operation of a propnet. Let us assume that there is an input marking and a base marking. From these, we can compute a view marking, as we have just seen. Importantly, this includes the inputs to the transitions of the propnet. On the next step, a new input marking is imposed from without. However, the new base marking is determined by the transitions. If the inputs to the transitions are true, then the outputs of the transitions are true on the next step. In effect, a transition as a mechanism for controlling the flow of information from one step to the next. It is a 1-step delay, a flip-flop in digital circuitry.

As an example, once again consider the propnet in Fig. 9.3 with the marking illustrated in Fig. 9.4. Now, let's move on to the next step. Suppose the input marking for the second step is the same as the first, i.e., *a* is true and *b* is false. What is *s* on this step? Since *s* is the output of a transition, its value on this step is the same as the value of that transition's input on the preceding step. In this case, the transition's input was false on the preceding step, and so *s* is false on this new step. As before, we can compute the view marking corresponding to the input marking and

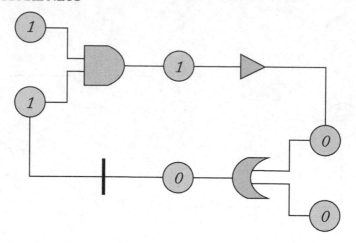

Figure 9.4: One marking for a propositional net.

this new base marking. See Fig. 9.5. In this case, since the second input to the and-gate is false, p is false and q is true and, therefore, r is true as well.

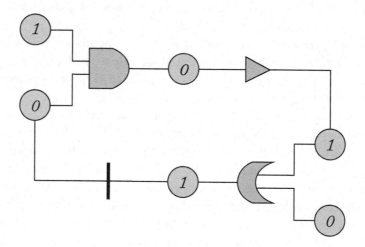

Figure 9.5: Another marking.

From this input marking and the new base marking, we can compute a new view marking. And we can then repeat. In this case, the value of proposition s will go on alternating between true and false so long as input a is true and input b is false. If input a ever becomes false, it will stop alternating. However, the alternation will begin again as soon as it is set to true again.

9.3   GAMES AS PROPOSITIONAL NETS

Propnets are an alternative to GDL for expressing the dynamics of games. With a few additional provisions, it is possible to convert any GDL game description into a propositional net with the same dynamics.

As an example of this, consider the simple game described below. There is just one role. There is just one base proposition, and there are two actions. The two actions are always legal. The player gets 100 points in any state in which s is true; otherwise, the player gets 0 points. The game ends if q ever becomes true. (Note that termination here is defined indirectly in terms of the action of the player. Properly, it should be defined entirely in terms of the state of the game and nothing else. This shortcoming can be fixed in various ways, but doing so would complicate the example.)

```
role(white)
base(s)
input(white,a)
input(white,b)
legal(white,a)
legal(white,b)
p :- does(white,a) & true(s)
q :- ~p
r :- q
r :- does(white,b)
next(s) :- r
goal(white,100) :- true(s)
goal(white,0) :- ~true(s)
terminal :- q
```

Now, let's build a propnet for this game. The base propositions in the propnet consist of the propositions defined by the base relation in the game description (viz. s). The input propositions correspond to the actions defined by the input relation in the game description (viz. a and b).

We use the next relation to capture the dynamics of the game. Starting with the base and input propositions, we add links for each rule (using inverters for negations, and-gates for multiple conditions, and or-gates for multiple rules). In so doing, we augment the propnet with additional view propositions as necessary. The result is the propnet shown in Fig. 9.3.

We model terminal in the propnet by adding a special node for termination. If necessary, we can extend the propnet to include new view propositions, as we do for next. In this case, terminal corresponds exactly to q, so we could just use q as our terminal node. However, for the sake of clarity, we can add a new node t and insert a connection from q to t. See Fig. 9.6. Note that we use a one-input and-gate here. We could equally well use a two-input and-gate with both inputs supplied by q, but this is simpler. The behavior is the same.

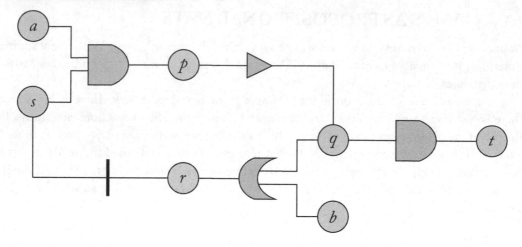

Figure 9.6: Propnet with terminal node.

Rewards are handled analogously. We create a new node for each reward value, and we use the definitions for these values to extend the propnet further. In this case, goal(white,100) corresponds exactly to s, so we do not need to add a new node for goal(white,100). However, for clarity, in our example, we have added one node for each of the two goal values. See Fig. 9.7.

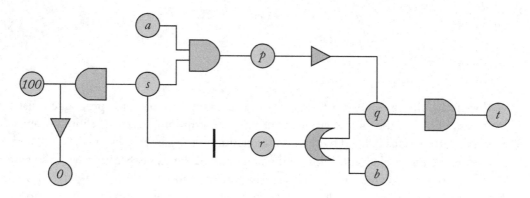

Figure 9.7: Propnet with goal nodes.

Legality is the trickiest part. There are various ways of doing this. The simplest method conceptually is to add one legality propositions for each possible action and extend the propnet to say when these nodes are true. In this case, we add two new propositions la and lb, corresponding to actions a and b. In this case they are always true. To model this, we add a self-loop - a transition for each of these legality propositions—and we initialize to true. Because of the dynamics of transitions, these propositions will remain true indefinitely.

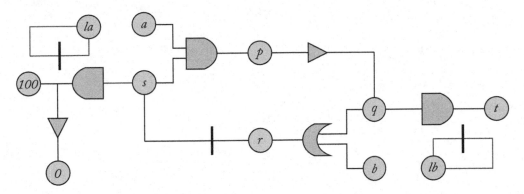

Figure 9.8: Propnet with legality nodes.

That's it. Once this is done, we have a propnet that reflects the game described in the given GDL. In the next chapter, we discuss how to use this propnet to play games.

CHAPTER 10

General Game Playing With Propnets

10.1 INTRODUCTION

As we saw in Chapter 4, a game player is typically implemented as a web service that receives messages from a Game Manager and replies appropriately. Building a player means writing event handlers for the different types of messages in the GGP communication protocol using subroutines for processing game descriptions and match data. Our job in building a player is to use the available subroutines to implement event handlers for the various GGP messages.

Building a game player using a propnet is essentially the same. In fact, all of the methods we have described so far still apply. The main difference is that we replace the subroutines that work directly on game descriptions with subroutines that operate on propnets. For example, in place of the `findlegals` subroutine we used earlier, we now use a `proplegals` subroutine that computes legal actions using the propnet; in place of the `findterminalp` subroutine we used earlier, we now use a `propterminalp` subroutine that computes whether or not a state is terminal using the propnet; and so forth.

We begin this chapter by presenting a representation of propnets as data structures. We then define the basic subroutines for marking propnets and reading marks on propnets. Then we use these subroutines to build up the General Game Playing subroutines needed by our players (e.g., `proplegals`, `propterminalp`, and so forth).

10.2 PROPOSITIONAL NETS AS DATA STRUCTURES

Before we can define methods for processing propnets, we need a representation of those propnets as data structures. In the simple approach taken here, we represent each propnet component as a structured object with various components depending on the type of the object. The connectivity of the propnet is captured by values of these components.

A propnet as a whole is represented by an object of the form shown below. The type is `propnet`. The `roles` component is a sequence of the roles in the game. The `inputs` component is a sequence of sequences of input nodes, one sequence for each role; the `bases` component is a sequence of the base nodes in the game; and the `view` component is a sequence of the view nodes in the game. The `legals` component is a sequence of sequences of legal nodes, one sequence for each role; the `rewards` component is a sequence of sequences of reward nodes; and the terminal component is the terminal node of the game.

```
{type:'propnet',
 roles:roles, inputs:inputs, bases:bases, views:views
 legals:legals, rewards:rewards, terminal:terminal}
```

An input node is a structured object of the form shown below. The `type` component is `input`; the `name` is the GDL representation of the input; and the mark is an indication of whether the node is true or false in the current state (as determined by the input marking current at each point in time).

```
{type:'input',name:name,mark:boolean }
```

A base node is a structured object of the form shown below. The `type` component is `base`; the `name` is the GDL representation of the node; the source is the transition that leads to the base node; and the mark is an indication of whether the node is true or false in the current state (as determined by the base marking current at each point in time).

```
{type:'base',name:name,source:transition,mark:boolean}
```

A view node is a structured object of the form shown below. The `type` component is `view`; the `name` is the GDL representation of the node; and the source is the connective that leads to the view node.

```
{type:'view',name:name,source:connective}
```

A negation is a structured object of the form shown below. The `type` component is `negation`, and the source is the proposition that leads to the connective.

```
{type:'negation',source:source}
```

A conjunction is a structured object of the form shown below. The `type` component is `conjunction`, and the sources component is a sequence of propositions that leads to the connective.

```
{type:'conjunction',source:sources}
```

A disjunction is a structured object of the form shown below. The `type` component is `disjunction`, and the sources component is a sequence of propositions that leads to the connective.

```
{type:'disjunction',sources:sources}
```

A transition is a structured object of the form shown below. The `type` component is `transitin`, and the source is the proposition that leads to the transition.

```
{type:'transition',source:source}
```

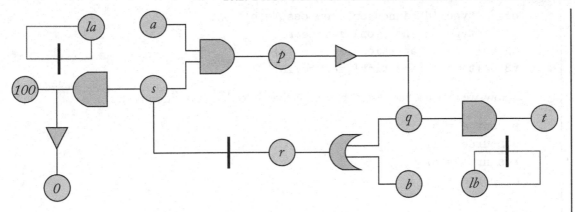

Figure 10.1: Propnet for a simple game.

As an example, consider the propnet presented in the preceding chapter. For convenience, it appears here again as Figure 10.1. In what follows, we see how to assemble this propnet manually. In a general game player, it would be automatically generated from the GDL game description.

First of all, we create input, base, and view nodes for the propositions in the propnet. Since we do not yet have data structures for our connectives, we initialize the sources of our propositions with null.

```
a = {type:'input',name:'a'} b = {type:'input',name:'b'}
p = {type:'view',name:'p',source:null,mark:false}
q = {type:'view',name:'q',source:null,mark:false}
r = {type:'view',name:'r',source:null,mark:false}
s = {type:'base',name:'s',source:null,mark:false}

la = {type:'base',name:'la',source:null,mark:false}
lb = {type:'base',name:'lb',source:null,mark:false}
g100 = {type:'view',name:'100',source:null,mark:false}
g0 = {type:'view',name:'0',source:null,mark:false}
t = {type:'view',name:'terminal',source:null,mark:false}
```

Next, we create connectives and insert the propositions as components.

```
a1 = {type:'conjunction',sources:[a,s]}
a2 = {type:'conjunction',sources:[q,q]}
a3 = {type:'conjunction',sources:[s,s]}
i1 = {type:'negation',source:p}
i2 = {type:'negation',source:g100}
```

```
o1 = {type:'disjunction',sources:[q,b]}
t1 = {type:'transition',source:r}
t2 = {type:'transition',source:la}
t3 = {type:'transition',source:lb}
```

Now, we go back and insert the connectives into the proposition nodes.

```
p.source = a1
q.source = i1
r.source = o1
s.source = t1

la.source = t2
lb.source = t3
g100.source = a3
g0.source   = i2
t.source = a2
```

Finally, we insert all of these things into the components of mypropnet.

```
mypropnet =
 {type:'propnet',
  inputs:[a,b],
  bases:[s, la, lb],
  views:[p,q,r],
  rewards:[g100,g0],
  terminal:t}
```

Once we have created our propnet, either manually or automatically from the GDL for a game, we can access the components by simply naming the propnet and one of its components and extracting the desired component. For example, to get a sequence of the base propositions in mypropnet, we would ask for mypropnet.base; and to get the source of proposition r, we would ask for r.source.

10.3   MARKING AND READING PROPOSITIONAL NETS

In our approach to General Game Playing using propnets, we use input markings in place of moves and base markings in place of states. In order to compute the various attributes of a game on a given step, we typically mark the input propositions and base propositions of the propnet and then compute the corresponding view marking. We then read the view marking to compute the desired attributes. In this section, we discuss the details of marking and reading propnets, and in the next section we show how they are used.

There are two different approaches to marking and reading propnets. In the *forward propagation* approach, we mark propositions and then propagate the values to compute the values of view propositions dependent on those marks. To answer questions, we then just read the values of the relevant view propositions. In the *backward reasoning* approach, we mark our base and input propositions but do not propagate. Instead, when we need a value for a view proposition, we work backward from the desired view proposition to determine that value. The forward method saves redundant computation. The backward methods is a little simpler and not much slower. We show the backward method here.

Marking propositions is easy. See the subroutines below. We simply iterate through the base propositions or input propositions marking those nodes with values from the input or base vectors of booleans.

```
function markbases (vector,propnet)
 {var props = propnet.bases;
  for (var i=0; i<props.length; i++)
      {props[i].mark = vector[i]};
  return true}

function markactions (vector,propnet)
 {var props = propnet.actions;
  for (var i=0; i<props.length; i++)
      {props[i].mark = vector[i]};
  return true}

function clearpropnet (propnet)
 {var props = propnet.bases;
  for (var i=0; i<props.length; i++)
      {props[i].mark = false};
  return true}
```

Computing the values of view propositions is accomplished by working backwards from the propositions of interests. If the proposition is an input or base proposition, we simply return the mark on that proposition. In the case of a view proposition, we compute the values of propositions for the inputs to the connective feeding the view proposition and combine those values in accordance with the type of the connective.

```
function propmarkp (p)
 {if (p.type=='base') {return p.mark};
  if (p.type=='input') {return p.mark};
  if (p.type=='view') {return propmarkp(p.source)};
  if (p.type=='negation') {return propmarknegation(p)};
```

```
    if (p.type=='conjunction') {return propmarkconjunction(p)};
    if (p.type=='disjunction') {return propmarkdisjunction(p)};
    return false}

function propmarknegation (p)
 {return !propmarkp(p.source)}

function propmarkconjunction (p)
 {var sources = p.sources;
  for (var i=0; i<sources.length; i++)
      {if (!propmarkp(sources[i])) {return false}};
  return true}

function propmarkdisjunction (p)
 {var sources = p.sources;
  for (var i=0; i<sources.length; i++)
      {if (propmarkp(sources[i])) {return true}};
  return false}
```

10.4   COMPUTING GAME PLAYING BASICS

Now that we have some tools for marking and reading propnets, let's see how we can use them to define the basic subroutines used in the game playing methods we defined in previous chapters.

In order to compute the legal actions for a role in a given state using a propnet, we first mark the base propositions of the propnet using the information in the given state. We then check each of the legality nodes for the given role in the propnet. If the node is true, we put the corresponding input node on the list. When we are done, we return the list of input nodes that we have accumulated in this process. (Note that, to return one of these actions to a game manager, we would need to extract the name component from the corresponding input node.)

```
function proplegals (role,state,propnet)
 {markbases(state,propnet);
  var roles = propnet.roles;
  var legals = seq();
  for (var i=0; i<roles.length; i++)
      {if (role==roles[i]) {legals = propnet.legals[i]; break}};
  var actions = seq();
  for (var i=0; i<legals.length; i++)
      {if (propmarkp(legals[i]))
          {actions[actions.length]=legals[i]}};
  return actions}
```

The propnext subroutine shown below computes the next state for a given move and a given state using a given propnet. (Note that the move here is assumed to be an input marking, not just a sequence of actions. To simulate a move supplied by a game manager, we would need to convert to an input marking before using propnext.) In executing the propnext subroutine, we first mark the input propositions using the given move, and we mark the base propositions of the propnet using the information in the given state. We then check each of the base propositions in the propnet, collecting those that are true in the next state.

```
function propnext (move,state,propnet)
 {markactions(move,propnet);
  markbases(state,propnet);
  var bases = propnet.bases;
  var nexts = seq();
  for (var i=0; i<bases.length; i++)
      {nexts[i] = propmarkp(bases[i].source.source)};
  return nexts}
```

To compute the reward for a given role in a given state using a propnet, we first mark the base propositions as before. We then check each of the reward propositions for the given role until we find one that is true.

```
function propreward (role,state,propnet)
 {markbases(state,propnet);
  var roles = propnet.roles;
  var rewards = seq();
  for (var i=0; i<roles.length; i++)
      {if (role==roles[i]) {rewards = propnet.rewards[i]; break}};
  for (var i=0; i<rewards.length; i++)
      {if (propmarkp(rewards[i])) {return rewards[i].name}};
  return 0}
```

Computing whether or not a state is terminal is particularly easy. We mark the base propositions as before and check the terminal node for the propnet.

```
function propterminalp (state,propnet)
 {markbases(state,propnet);
  return propmarkp(propnet.terminal)}
```

CHAPTER 11

Factoring

11.1 INTRODUCTION

A *compound game* is a single game consisting of two or more individual games. The state of a compound game is a composition of the states of the individual games. On each step of a compound game, the players perform actions in each of the individual games. A compound game is over when either one or all of the individual games are over (depending on the type of compound game).

As an example, consider Hodgepodge. One state of the game is illustrated below. Hodgepodge is a combination of Chess and Go. On each step, a player makes a move on each board.

 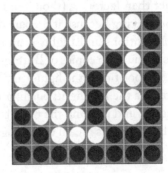

Figure 11.1: One state in Hodgepodge.

Using the techniques we have seen thus far, compound games can be quite expensive to play. Unless a player recognizes that there are independent subgames, it is likely to search a game tree that is far larger than it needs to be. If one subgame has branching factor a and a second has branching factor b, then the branching factor of the joint game is $a \times b$, and the fringe of the game tree at depth d is likely to be something like $(a \times b)^d$. This is wasteful if the two subgames are independent. In that case, there are two trees—one with branching factor a and one with branching factor b—and the total size of the fringe of these trees at depth d should be only $a^d + b^d$.

Factoring is the process of discovering independent games inside of larger games. Once discovered, game players can use these factors to play the individual games independently of each other and thus cut down on the combinatoric cost of playing such games. It turns out that it is often easier to discover independent subgames using the propnet representation of games rather than the original GDL.

In this chapter, we look at some elementary techniques for factoring games using propnets. In Section 11.2, we talk about discovering factors with completely independent subgames. In Section 11.3, we talk about factoring with interdependent goals and rewards; and, in Section 11.4, we talk about factoring with interdependent actions. Finally, in Section 11.5, we discuss conditional factors, i.e., factors that appear only in certain states of games.

11.2 COMPOUND GAMES WITH INDEPENDENT SUBGAMES

We begin our discussion of factoring with the simple case of compound games consisting of multiple completely independent subgames only one of which is relevant to the overall game. Without factoring, a player is likely to consider actions in all subgames when only one of the subgames matters. This is admittedly a very special case. It does not arise often, and it can be solved by means other than factoring (though not by the methods we have seen thus far). Nevertheless, it is worth considering because it prepares us for factoring more complicated games.

Multiple Buttons and Lights is an example of a game of this sort. See below. The overall game consists of multiple copies of Buttons and Lights. In each copy of Buttons and Lights, there are three base propositions (the lights) and three actions (the buttons). Pushing the first button in each group toggles the first light; pushing the second button in each group interchanges the first and second lights; and pushing the third button in each group interchanges the second and third lights. Initially, the lights are all off. The goal is to turn on all of the lights in the middle group. (The settings of the other lights are irrelevant.)

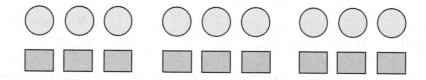

Figure 11.2: Multiple buttons and lights.

The GDL for this game is shown below. There is just one role. There are nine base propositions, and there are nine actions. All actions are legal at all times. Each $a(i)$ toggles the corresponding $p(i)$. Each $b(i)$ interchanges the values of $p(i)$ to $q(i)$. And each $c(i)$ interchanges the values of $q(i)$ and $r(i)$. The game ends when $p(2)$ and $q(2)$ and $r(2)$ are true. In this case, the player gets 100 points. (In violation of our rule about finiteness, there is no step counter, so this game could, in principle, run forever.)

```
role(white)

base(p(X)) :- index(X)
base(q(X)) :- index(X)
base(r(X)) :- index(X)

input(white,a(X)) :- index(X)
input(white,b(X)) :- index(X)
input(white,c(X)) :- index(X)

index(1)
index(2)
index(3)

legal(a(X)) :- index(X)
legal(b(X)) :- index(X)
legal(c(X)) :- index(X)

next(p(X)) :- does(white,a(X)) & ~true(p(X))
next(p(X)) :- does(white,b(X)) &  true(q(X))
next(p(X)) :- does(white,c(X)) &  true(p(X))
next(q(X)) :- does(white,a(X)) &  true(q(X))
next(q(X)) :- does(white,b(X)) &  true(p(X))
next(q(X)) :- does(white,c(X)) &  true(r(X))
next(r(X)) :- does(white,a(X)) &  true(r(X))
next(r(X)) :- does(white,b(X)) &  true(r(X))
next(r(X)) :- does(white,c(X)) &  true(q(X))

goal(white,100) :- terminal
goal(white,0) :- ~terminal

terminal :- true(p(2)) & true(q(2)) & true(r(2))
```

Figure 11.3 shows the propnet for Multiple Buttons and Lights. There are three disjoint parts of the propnet—one portion for the first group of buttons and lights, a second portion for the second group, and a third portion for the third group. Note that the goal and termination conditions are based entirely on the lights in the second group.

Looking at the propnet for this game, it is easy to see that the game has a very special structure. The propnet consists of three completely disconnected subnets, one for each subgame.

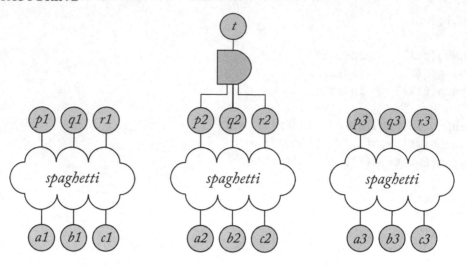

Figure 11.3: Propnet for multiple buttons and lights.

Finding factors in situations like this is easy—it can be computed in time that is polynomial in the size of the propnet.

Note that this technique can be applied equally well to multi-player games. Consider, for example, Multiple Tic-Tac-Toe, i.e., three games of Tic-Tac-Toe glued together in which only the middle game matters. See below.

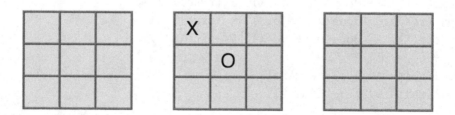

Figure 11.4: Multiple Tic-Tac-Toe.

The propnet for Multiple Tic-Tac-Toe is similar to the propnet for Multiple Buttons and Lights, and it is possible to find the structure for Multiple Tic-Tac-Toe just as easily as for Multiple Buttons and Lights.

11.3 COMPOUND GAMES WITH INTERDEPENDENT TERMINATION

In this section, we consider *compound games with interdependent termination*. As with the games discussed in the preceding section, actions are partitioned over distinct subgames and there are no incoming connections between those subgames. The main difference is that the termination condition for the overall game can depend on more than one and perhaps all of the subgames.

In games of this sort, the termination of the overall game can be defined as any boolean combination of conditions in the individual subgames. In the case where the combination is a disjunction, the game is said to have *disjunctive termination*. In the case where the combination is a conjunction, the game is said to have *conjunctive termination*.

As a simple example of a factorable game with disjunctive goals, consider Best Buttons and Lights. In Multiple Buttons and Lights, as defined in the preceding section, only one group of buttons and lights matters. In Best Buttons and Lights, the compound game terminates whenever *any* group terminates. This gives the player the freedom to play whichever subgame it likes and rest assured that, if it succeeds on that subgame, it succeeds on the overall game with the same score.

Figure 11.5 presents a sketch of the propnet for this game.

The good news is that we can extend the techniques discussed in the preceding section to this case. Let's consider the disjunctive case. If the connective leading to a termination is a disjunction with its inputs in turn supplied by nodes in different subgames, then we simply cut off that node and inputs to the or-gate termination nodes for the overall game. We repeat this process so long as we encounter only disjunctions. If the game is truly disjunctive, this will lead to a separable propnet. See Fig. 11.6.

If this process succeeds in factoring the propnet, then the player simply picks one of the subgames and proceeds as with the case of independent subgames. Alternatively and better, the player tries playing all of the subgames and picks the one with the best score. Or at least that is the basic idea.

Unfortunately, it is not quite that simple. There is a problem that does not arise in the case of completely independent subgames. The player may choose a subgame, find a winning strategy, and begin executing that strategy. Unfortunately, in the course of execution on the chosen subgame, one of the other subgames may terminate, terminating the game as a whole before the strategy in the chosen subgame is complete. This can lead to a lower score than the player might have expected.

The solution to this problem is to check each of the subgames for termination when no actions are played. We take the shortest time period and play each of the other subgames with that as step limit and take the one that provides the best result. For the subgame with the shortest termination, if there is no other game with that step limit, we try the next shortest step limit.

Although in this approach, the player searches all of the subgames more than once, this is usually a lot less expensive that searching the game tree for the unfactored game because it is not

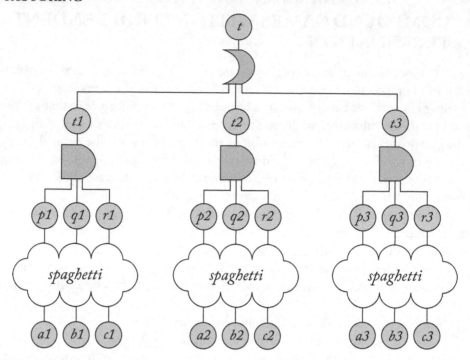

Figure 11.5: Propnet for best buttons and lights.

cross multiplying the branching factors of the independent games as it would if it did not use the game's factors.

11.4 COMPOUND GAMES WITH INTERDEPENDENT ACTIONS

In this section, we look at compound games with interdependent actions. By interdependence of actions, we mean that those actions have effects on more than one of the subgames of the compound game.

On first blush, it might seem that games of this sort are not factorable. However, this is not necessarily true. Under certain circumstances, even with interdependent actions, it is possible to identify factors and use those factors to search the game tree more efficiently than without considering these factors.

As an example, consider the game of Joint Buttons and Lights. See the illustration in Fig. 11.7. As with the other Buttons and Lights variants that we have seen, the lights in this game are organized into groups of three. However, unlike the previous variants, buttons are not associated with specific groups. Instead, each button has effects on all groups. Button *aaa* toggles

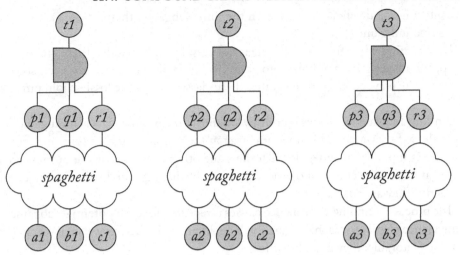

Figure 11.6: Modified propnet for best buttons and lights.

the first light in the first group *and* the first light in the second group *and* the first light in the third group. Button *aab* toggles the first light in the first group and the first light in the second group and interchanges the value of the first light and the second light in the third group. And so forth.

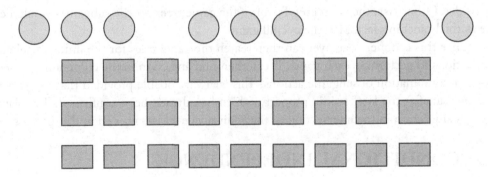

Figure 11.7: Joint buttons and lights.

Although all of the buttons in this game affect all of the subgames, the game is still factorable with just three actions per factor. The reason is that there is one button in the compound game for each combination of actions in the other two subgames. Thus, the game trees for each subgame can be searched independently of each other and the actions chosen can be reassembled into overall actions of the compound game. For example, if action *a* is chosen in the first and second subgames and action c is chosen in the third subgame, then action aac can be performed in the compound game.

Recognizing when this can be done is not easy, but it is doable. There are two steps. In the first step, the player groups actions into equivalence classes of actions for each subgame. In the second step, the player determines whether these classes satisfy the lossless join property (defined below).

Finding equivalence classes is done by looking at the propnet for each subgame. Two actions are equivalent if they are used in the same way in the propnet for that subgame. For example, if each action is input to an and-gate with the same other input and if the outputs are connected by an or-gate, then the effects of one action cannot be distinguished from the effects of the other action; hence they are equivalent.

The process of finding equivalence classes is repeated for each potential subgame. In general, the equivalence classes for each subgame will be different. In fact, as we shall see, in order for the game to be factorable, they must be different.

Once a player has equivalence classes for each potential subgame, it then checks whether those equivalence classes are independent of each other. The criterion is simple. Each equivalence class in one potential subgame must have a non-empty intersection with each equivalence class of every other potential subgame. If this is true, then the partitions pass the lossless join criterion.

If a player finds equivalence classes for two potential subgames and they pass this lossless join test, then the player can factor the game into subgames. In order to benefit from the factoring, the player must modify each propnet so that the individual actions are replaced with the equivalence classes of which they are members. This cuts down on the number of possible actions to consider. Otherwise, the branching factor of the game trees for the subgames would be just as large as the branching factor for the overall game.

Once this is done, the player can then search the game trees for the different subgames to select actions to perform in each game. It can then find an action in the compound game for the particular combination of subgame actions. This is always doable provided that the partitioning of actions satisfies the lossless join property, which has already been confirmed. The player then performs any action in the intersection of the equivalence classes chosen in this process.

11.5   CONDITIONAL INDEPENDENCE

We now consider the class of games that, over time, become factorable. For example, a game might not initially be separable into independent games; but it may, after entering a certain state, become separable.

Consider the following game, called Joint Tic-Tac-Toe - two games of Tic-Tac-Toe connected by a single square that connects the two. The goal of the game for a player is to get a line, a row, column or diagonal of that player's mark, with at least two of the marks residing in a specific Tic-Tac-Toe domain. Diagonals through the middle square do not count. On each turn, the player in control can place two marks, either one in each distinct Tic-Tac-Toe domain or one mark in a Tic-Tac-Toe domain and one in the center square. Suppose that the state of the game is as follows.

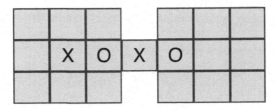

Figure 11.8: Joint Tic-Tac-Toe.

Once it is not possible for either player to achieve a row utilizing the center square, the only possible solutions lie in the domains of the Tic-Tac-Toe games that are joined by the center square. The states of the two Tic-Tac-Toe games can be considered independently to find the remaining optimal moves for the duration of the game. Only the game trees for each Tic-Tac-Toe game, modulo the center square, need to be searched to determine the remaining optimal moves. Given the current state, the game can be factored into two independent sub-games. However, from the initial state, this game cannot be factored into independent games, because the shared middle square intertwines the two domains as it is relevant to the satisfaction of goals in both sub-games.

While this game is not factorable in general, it is factorable contingent upon entering a state in which no row through the middle is possible for either player. We define the game as being contingently factorable, since it reduces to independent simultaneous sub-games, given that it enters a certain state. The raw computational benefit acquired from recognizing contingent factors is less than that of recognizing initial factors, because it requires that the game be in a specific state. However, the relative computational savings are still the same, because the number of accessible fringe nodes reduces to the sum of the remaining accessible fringe nodes of the individual games rather than the product.

At this point in time, there are no established techniques for discovering and exploiting cases of conditional independence. However, it seems likely that some of the techniques just discussed can be adapted to this case as well.

CHAPTER 12

Discovery of Heuristics

12.1 INTRODUCTION

The biggest open problem in General Game Playing is the discovery of game-specific pruning rules and/or evaluation functions. In many games, it is possible to find such heuristics. The challenge to find such heuristics without searching the entire game tree.

In this chapter, we look at just one technique of this sort. We start by defining latches and inhibitors and discuss how to find them. Then we show how the concepts work together in a useful technique for removing states from the game tree and ordering states within the game tree.

12.2 LATCHES

A *latch* is a proposition that, once it becomes true (or, respectively, false), remains true (or, respectively, false) no matter what the players do. A latch is not the same as a constant in that it may change value during the game; but, once the value changes, it keeps the new value for all remaining steps.

Almost all of the propositions in our usual description of Tic-Tac-Toe are latches. Once a square is marked, it keeps that mark for the rest of the game; and once a cell ceases to be blank, it can never again become blank. In other words, propositions like cell(1,2,x) and cell(2,3,o) and cell(3,3,b) are all latches.

Given a propnet, it is straightforward to find latches. We enumerate values for all base markings and check the value of the candidate proposition. If the value remains the same, then the proposition is a latch.

One feature of this approach is that we do need to consider arbitrary action histories, only possible combinations of actions. The downside is that we still need to cycle through all possible states of the propnet as well. And thus the technique is prohibitively expensive in general.

A partial solution to this problem is to restrict one's attention to those base and input propositions that determine the proposition being checked. A set of base and input propositions determine all and only the propositions in the propnet that are downstream from it, i.e., the propositions themselves or the outputs of connectives for which they are the inputs and so forth. In the worst case, this may be as bad as the full search just described, but in many cases it is much less expensive.

Note that the idea of latches can be extended to groups of propositions. While no individual proposition in the group is a latch, it is possible that the group as a whole might constitute a latch.

Consider, for example, a game in which one of p or q is always true. Then neither p nor q is a latch but their disjunction does form a latch, even if that disjunction is not a node in the propnet. Obviously, finding composite latches is more expensive than finding individual latches, but it may be helpful in some games.

12.3   INHIBITORS

A proposition p inhibits a proposition q if and only if p must be false to achieve q or retain q. In other words, whenever p is true in a state, q is false in the next state. When this occurs, we say that p is an inhibitor of q.

Some inhibitors can be detected by a simple scan of the propnet for a game. If every path from the base propositions and input propositions to q involve the negation of p, then p is an inhibitor of q.

12.4   DEAD STATE REMOVAL

Dead state removal is the removal of dead states from a game tree, i.e., states that cannot lead to satisfactory terminal states, i.e., those that can give a player an adequate score (e.g., 100 points, points above a desired threshold, or a plurality of points in a zero sum game). Pruning the game tree below dead states can save lots of computation.

In a propnet, if the truth (or falsity) of a proposition makes it a dead state, then we should strive to ensure that the that proposition never becomes true (or, respectively, false). So the technique could equally well be called dead proposition detection.

The trick in dead state removal is to find such states without actually searching the game tree starting with those states. Many ways of doing this have been proposed. In this section, we examine one such technique based on the concepts of latches and inhibitors.

As an example, consider a game with the following behavioral rule. White gets 100 points if both q and r are true. If we assume that this is the only way in which white can get 100 points, then both q and r are goal inhibitors.

```
goal(white,100) :- true(q) & true(r)
```

Now let's add in the rule shown below. If q ever becomes true, it remains true forever. In other words, q is a latch. Obviously, in a situation like this, we would like to ensure that q never becomes true or we lose the possibility of getting 100 points.

```
next(q) :- true(q)
```

Generalizing this gives us our rule for dead state removal / dead proposition setting—if a latch inhibits the goal in a game, then we should not consider states in which that proposition is set to true.

As an example of dead state removal in action, consider Untwisty Corridor. It is a single-player game. There are nine propositions: p and q1,..., q8. And there are four actions: a, b, c, and

d. q1 is true in the initial state, and the goal is to turn on q8. Actions a, b, and c all turn on p. If a q proposition is true, then action d turns on the next in sequence, provided that p is false; otherwise it does nothing. There is a step counter that terminates the game after seven steps.

The game tree for Untwisty Corridor has 349,525 states. However, most of those states are dead states. Recognizing that p is a latch and a goal inhibitor allows us to prune all of these dead states, leaving a total of 8 states to be considered, a considerable savings.

Note that this basic idea can be generalized in various ways. For example, we can look for latches that inhibit 90 point states and latches that inhibit 80 point states and so forth; and we can use our discoveries to order states depending on the values of the goals that are affected.

CHAPTER 13

Logic

13.1 INTRODUCTION

In the last few chapters we looked at propositional nets as an alternative to the Game Description Language for encoding games. In this and the following chapters we return to GDL as the typical language in which the rules of games are communicated in general game playing.

We first look at the task of computing logical consequences from a logic program. This provides the background for implementing the basic functionality of a general game player to generate legal moves, compute state update, and decide termination from a GDL game description (see Chapter 4).

As we will see, the ability to draw inferences from a logic program is also needed for converting a GDL input into a propnet or any other data structure that is more efficient to compute with. We will show how logic is used to find structure in GDL games such as symmetries, which help improve the performance of a general game-playing system. We will look into how single-player games can be solved in logic and how logic can be used to aid decision making through the automatic construction of a goal-oriented evaluation function.

The basic type of inference we are concerned with can be formulated as *queries* that ask whether a literal L, or a conjunction of literals L_1 & L_2 & ... & L_n, follows from a set of clauses. Often a query contains variables, and then we are interested in obtaining values for these variables under which the query becomes true.

Suppose, for instance, that we want to infer a legal move for a player, say white, in a particular state of a game. We can formulate this as the query `legal(white,L)`, which means to determine for which L, if any, this is a logical consequence of the given rules.

Let's use as an example the GDL description of a 2-player game known as Nim. Players take turns removing one or more objects from one of several heaps. The game rules 1–10 below define a legal move as reducing the size of a selected heap to a smaller value (Fig. 13.1). The facts in lines 12–15 encode a randomly chosen current state.

13.2 UNIFICATION

The most basic step in computing a query to a logic program is called *unification*. It means the process of replacing variables so that two logical expressions become similar. This is needed to determine which rule from the program could provide an answer to an atomic query such as `legal(white,L)`.

```
 1   legal(P,reduce(X,N)) :-
 2      true(control(P)) & true(heap(X,M)) & smaller(N,M)
 3
 4   smaller(X,Y) :- succ(X,Y)
 5   smaller(X,Y) :- succ(X,Z) & smaller(Z,Y)
 6   succ(0,1)
 7   succ(1,2)
 8   succ(2,3)
 9   succ(3,4)
10   succ(4,5)
11
12   true(heap(a,2))
13   true(heap(b,0))
14   true(heap(c,5))
15   true(control(white))
```

Figure 13.1: A collection of rules from the game Nim along with facts encoding a given state.

Generally, speaking, a program fact A, or a program rule $A : -B_1$ & ...& B_m, can only provide an answer to a query atom L if it is possible to replace the variables occurring in L and A in such a way that the two atoms become similar.

Definition 13.1   A *substitution* is a finite set of replacements $\{x_1/t_1, \ldots, x_n/t_n\}$ such that

- $n \geq 0$;

- $x_1, \ldots, x_n$ are pairwise distinct variables; and

- $t_1, \ldots t_n$ are terms (which may or may not contain variables).

The result $\text{SUBST}(E, \sigma)$ of applying a substitution $\sigma = \{x_1/t_1, \ldots x_n/t_n\}$ to an expression E is obtained by simultaneously replacing in E every x_i by its replacement t_i.

Definition 13.2   Two expressions E_1 and E_2 are *unifiable* if we can find a substitution σ such that $\text{SUBST}(E_1, \sigma) = \text{SUBST}(E_2, \sigma)$. Any such σ is called a *unifier* of E_1 and E_2.

Recalling the example from above, the atomic query legal(white,L) and the head, legal(P,reduce(X,N)), of the first rule in Fig. 13.1, are unifiable. Indeed, there are many different unifiers for the two, including all of the following.

$\sigma_1 = \{$P/white,L/reduce(a,1),X/a,N/1$\}$:
  SUBST(legal(white,L), σ_1)   = legal(white,reduce(a,1))
  SUBST(legal(P,reduce(X,N)), σ_1)   = legal(white,reduce(a,1))

$\sigma_2 = \{$P/white,L/reduce(X,X),N/X$\}$:
  SUBST(legal(white,L), σ_2)   = legal(white,reduce(X,X))
  SUBST(legal(P,reduce(X,N)), σ_2)   = legal(white,reduce(X,X))

$\theta = \{$P/white,L/reduce(X,N)$\}$:
  SUBST(legal(white,L), θ)   = legal(white,reduce(X,N))
  SUBST(legal(P,reduce(X,N)), θ)   = legal(white,reduce(X,N))

Comparing the three substitutions, the last one, θ, appears to be the most general way of unifying our two atoms: While variable L is assigned a move of the form reduce(X,N) in all three cases, unifier σ_1 additionally fixes specific values for X and N whereas σ_2 binds the two variables together. But neither is necessary to unify legal(white,L) and legal(P,reduce(X,N)). This leads to Definition 13.3 below, which refers to the *composition* of two substitutions $\theta_1 \circ \theta_2$ as a substitution that, for any expression E, satisfies

$$\text{SUBST}(E, \theta_1 \circ \theta_2) = \text{SUBST}(\text{SUBST}(E, \theta_1), \theta_2)$$

Definition 13.3   A unifier θ is *more general* than a unifier σ if there is a third substitution τ such that $\theta \circ \tau = \sigma$.

A *most general unifier* for expressions E_1 and E_2 is one that is more general than any other unifier of the two expressions.

It follows that indeed our unifier θ from above is more general than both σ_1 and σ_2, as can be seen from

$$\sigma_1 = \theta \circ \{\text{X/a,N/1}\}$$
$$\sigma_2 = \theta \circ \{\text{N/X}\}$$

In fact, θ is a most general unifier for our example. Fortunately it is a known fact that whenever two atomic formulas are unifiable, then a most general unifier always exists and is unique up to variable permutation.

Moreover, most general unifiers can be easily computed. The recursive algorithm below takes as input two expressions and a partially computed unifier (which should be empty at the first function call). It is assumed that expressions as well as partial unifiers are encoded as lists.

For example, the function call

```
unify([legal,white,L],[legal,P,[reduce,X,N]],[])
```

returns the following most general unifier.

$$[[P,white],[L,[reduce,X,N]]]$$

The algorithm works by comparing the structure of the two expressions, argument by argument.

```
function unify (x,y,sigma)
 {if (x == y) {return sigma};
  if (is_variable(x))  {return unifyVariable(x,y,sigma)};
  if (is_variable(y)) {return unifyVariable(y,x,sigma)};
  if (! is_list(x))
     {if (is_list(y) || x != y) {return fail}
      else {return sigma}};
  if (! is_list(y)) {return fail};
  sigma = unify(head(x),head(y),sigma);
  if (sigma == fail) {return fail}
   else {return unify(tail(x),tail(y),sigma)}}

function unifyVariable(x,y,sigma)
 {if (var z = replacement_for(x,sigma)) {return unify(z,y,sigma)};
  if (var z = replacement_for(y,sigma)) {return unify(x,z,sigma)};
  if (x != y && occurs_in(x,y)) {return fail}
   else {return add_element([x,y],sigma)}}
```

Function replacement_for(x,sigma) is assumed to return the replacement for variable x according to unifier sigma if such a replacement exists; and 0 otherwise. Function occurs_in(x,y) should return true just in case variable x occurs anywhere inside the (possibly nested) list y.

13.3   DERIVATION STEPS (WITHOUT NEGATION)

Backward-chaining is the standard technique for query answering in clausal logic. For this we work our way backwards from the query, first by finding a rule that applies to the leading query element and then proving that the condition of the rule, i.e., its body, holds.

A clause is applicable to atoms that can be unified with its head. There is a small subtlety though. Suppose we had formulated the search for legal moves of our player as legal(white,X) instead of legal(white,L). We would then be unable to unify this query with the head of the clause shown below.

```
legal(P,reduce(X,N)) :- true(control(P)) & true(heap(X,M)) &  smaller(N,M)
```

The reason is that no substitution σ can possibly exist such that $\text{Subst}(\text{legal}(\text{white},X),$ $\sigma) = \text{Subst}(\text{legal}(P,\text{reduce}(X,N)),\sigma)$ because X cannot be equated with reduce(X,N).

But actually the scope of a variable never extends beyond the clause in which it appears. For this reason, each application of a rule should be preceded by generating a "fresh" copy in which the variables have been consistently replaced by new names that do not occur in the original query or any other step in the same derivation.

For a single derivation step we then proceed as follows. Let's first consider the case that neither the query nor any of the program clauses contains a negated literal.

Given: Clauses G (without negation)
        Query $L_1 \& L_2 \& \dots \& L_n$ (without negation), $n \geq 1$

Let: $A :- B_1 \& B_2 \& \dots \& B_m$ "fresh" variant of a clause in $G, m \geq 0$
        σ most general unifier of L_1 and A

Then: $L_1 \& L_2 \& \dots \& L_n \Rightarrow \text{Subst}(B_1 \& B_2 \& \dots \& B_m \& L_2 \& \dots \& L_n, \sigma)$

A derivation step thus replaces the leading element of the query by the body of a suitable clause and applies the necessary variable bindings to all of the new query.

Here is an example that uses the first clause of Fig. 13.1.

```
    legal(white,L)                          rule 1 with {P/white,L/reduce(X,N)}
⇒   true(control(white))
    & true(heap(X,M)) & smaller(N,M)
```

When the applied clause is a fact, i.e., a rule with empty body, then the respective query element is removed without replacement, as in the following derivation step.

```
    true(heap(X,M)) & smaller(N,M)   rule 12 with {X/a,M/2}
⇒   smaller(N,2)
```

13.4 DERIVATIONS

A complete derivation for a query is obtained by the repeated application of derivation steps until all subgoals have been resolved. If the original query includes variables, then the *computed answer* is determined by the variable bindings made along the way.

Definition 13.4 A series of derivation steps is called a *derivation*. A *successful* derivation is one that ends with the empty query, denoted as "□".

The *answer* computed by a successful derivation is obtained by computing the composition $\sigma_1 \circ \sigma_2 \circ \ldots \circ \sigma_k$ of the unifiers of each step and restricting the result to the variables in the original query.

An example is shown below, where again we use the clauses of Fig. 13.1.

	`legal(white,L)`	rule 1 with {P/white,L/reduce(X,N)}
$\Rightarrow$	`true(control(white))`	rule 15 with {}
	`& true(heap(X,M)) & smaller(N,M)`	
$\Rightarrow$	`true(heap(X,M)) & smaller(N,M)`	rule 12 with {X/a,M/2}
$\Rightarrow$	`smaller(N,2)`	copy of rule 4 with {X'/N,Y'/2}
$\Rightarrow$	`succ(N,2)`	rule 6 with {N/1}
$\Rightarrow$	□	

Answer: {L/reduce(a,1)}

Here is another successful derivation for the same query.

	`legal(white,L)`	rule 1 with {P/white,L/reduce(X,N)}
$\Rightarrow$	`true(control(white))`	rule 15 with {}
	`& true(heap(X,M)) & smaller(N,M)`	
$\Rightarrow$	`true(heap(X,M)) & smaller(N,M)`	rule 12 with {X/a,M/2}
$\Rightarrow$	`smaller(N,2)`	copy of rule 5 with {X'/N,Y'/2}
$\Rightarrow$	`succ(N,Z') & smaller(Z',2)`	rule 6 with {N/0,Z'/1}
$\Rightarrow$	`smaller(1,2)`	copy of rule 4 with {X''/1,Y''/2}
$\Rightarrow$	`succ(1,2)`	rule 7 with {}
$\Rightarrow$	□	

Answer: {L/reduce(a,0)}

Derivations aren't always successful.

Definition 13.5   A derivation *fails* if it leads to a query whose first element does not unify with the head of any available clause.

Here is an example of a derivation for the same query as above that leads to a dead-end.

	`legal(white,L)`	rule 1 with {P/white,L/reduce(X,N)}
$\Rightarrow$	`true(control(white))`	rule 15 with {}
	`& true(heap(X,M)) & smaller(N,M)`	
$\Rightarrow$	`true(heap(X,M)) & smaller(N,M)`	rule 13 with {X/b,M/0}
$\Rightarrow$	`smaller(N,0)`	copy of rule 4 with {X'/N,Y'/0}
$\Rightarrow$	`succ(N,0)`	
$\Rightarrow$	failure	

13.5 DERIVATION TREE SEARCH

Some of the basic functions from Chapter 4 for a general game player require to consider all possible derivations of a query. For instance, findlegals(*role*,*state*,*description*) is expected to deliver every computable answer to legal(*role*,M) for a given state and game description. *Backtracking* provides a systematic way to compute these: after having found the first successful or failed derivation, one goes back to the most recent choice where a different clause could have been applied to the query, until all choices have been exhausted.

We can draw a tree with the original query as the root to illustrate the search space of backward-chaining. Each computed answer then corresponds to one branch that ends with the empty query. Switching to a different example, consider the problem of computing *all* legal moves for a player in a given Tic-Tac-Toe position. Let's use the clauses listed below in Fig. 13.2. Specifically, the facts in lines 1–10 together encode the following game state, with white to move.

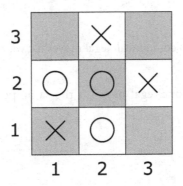

```
1    true(cell(1,1,x))
2    true(cell(1,2,o))
3    true(cell(1,3,b))
4    true(cell(2,1,o))
5    true(cell(2,2,o))
6    true(cell(2,3,x))
7    true(cell(3,1,b))
8    true(cell(3,2,x))
9    true(cell(3,3,b))
10   true(control(white))
11
12   legal(W,mark(X,Y)) :- true(cell(X,Y,b)) & true(control(W))
13   legal(white,noop) :- true(control(black))
14   legal(black,noop) :- true(control(white))
```

Figure 13.2: The rules for legality in Tic-Tac-Toe along with facts encoding a given position.

The tree depicted in Fig. 13.3 shows the three legal moves that can be computed for white in our example Tic-Tac-Toe position. The labels at the arcs indicate which clause has been selected. Backtracking can be implemented by a depth-first search through this tree (Fig. 13.3).

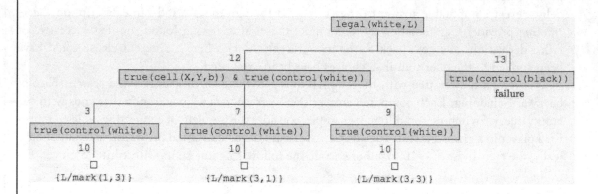

Figure 13.3: Derivation tree to compute all legal moves for white from the clauses in Fig. 13.2.

For black there is just one legal move in the same position, as the computation tree in Fig. 13.4 illustrates.

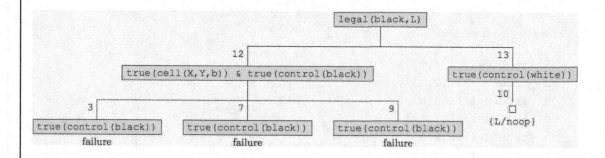

Figure 13.4: Derivation tree to compute all legal moves for black from the clauses in Fig. 13.2.

A correct implementation of a complete search through a derivation tree requires you to address the problem of potentially infinite derivations that may arise with some game descriptions. Even if a GDL describes a finite game, there is always the risk that a computation loops because of a recursive clause. Suppose, for example, the following rule were added to the game description of Nim in Fig. 13.1.

```
16   smaller(X,Y) :- smaller(X,Y)
```

This clause is obviously redundant but perfectly correct, both logically and syntactically. A straightforward implementation to compute derivations with this rule will eventually enter a non-terminating loop.

	`legal(white,L)`	rule 1 with {P/white,L/reduce(X,N)}
⇒	`true(control(white))`	rule 15 with {}
	`& true(heap(X,M)) & smaller(N,M)`	
⇒	`true(heap(X,M)) & smaller(N,M)`	rule 12 with {X/a,M/2}
⇒	`smaller(N,2)`	copy of rule 16 with {X'/N,Y'/2}
⇒	`smaller(N,2)`	...
⇒	...	

Although it may be unlikely that you encounter such a tricky game description in practice, it can be a good idea to add a simple loop check to avoid non-terminating computations. Note that potential loops may be less obvious to detect from the GDL itself than in our example if they are spread over several clauses.

Recursive clauses may even lead to non-terminating derivations in which the query grows forever without repetition. To see why, consider the following alternative formalization of clause 5 from Fig. 13.1.

```
5*  smaller(X,Y) :- smaller(Z,Y) & succ(X,Y)
```

Logically, this means the very same thing, but it may give rise to a non-looping, infinite derivation as follows.

	`legal(white,L)`	rule 1 with {P/white,L/reduce(X,N)}
⇒	`true(control(white))`	rule 15 with {}
	`& true(heap(X,M)) & smaller(N,M)`	
⇒	`true(heap(X,M)) & smaller(N,M)`	rule 12 with {X/a,M/2}
⇒	`smaller(N,2)`	copy of rule 5* with {X'/N,Y'/2}
⇒	`smaller(Z',2) & succ(N,Z')`	copy of rule 5* with {X''/Z',Y''/2}
⇒	`smaller(Z'',2)`	
	`& succ(Z',2) & succ(N,Z')`	...
⇒	...	

This problem can be avoided simply by reordering the atoms in a rule in such a way that recursive calls always come last.

13.6 HANDLING NEGATION

Let's now generalize our derivation procedure to queries and programs that include negative literals. A subgoal of the form ~A can be treated according to a principle known as *negation-by-failure*.

It says that $\sim A$ is true if A itself cannot be derived. This is justified by our use of the *minimal* model as the semantics of a set of GDL clauses.

For a single derivation step in the general case we thus proceed as follows.

Given: Clauses G
         Query $L_1 \& L_2 \& \ldots \& L_n$ $(n \geq 1)$

Case: $L_1 = A$ positive literal
         • proceed as before

Case: $L_1 = \sim A$ negative literal without variables
         • if all derivations for A fail: $L_1 \& L_2 \& \ldots \& L_n \Rightarrow L_2 \& \ldots \& L_n$
         • if a derivation for A succeeds: $L_1 \& L_2 \& \ldots \& L_n \Rightarrow$ failure

Recall, for example, one of the termination condition in Tic-Tac-Toe.

```
terminal :- ~open
open     :- true(cell(M,N,b))
```

As long as there is at least one blank cell in the current state, the sub-goal `true(cell(M,N,b))` has a successful derivation, and hence so has the atom `open`. Consequently, the negative sub-goal `~open` fails unless all cells have been marked. The query `terminal` therefore fails too, provided that no other terminating conditions hold that allow it to be derived.

But when a state is reached in which all cells have been marked, then `true(cell(M,N,b))` can no longer be derived, and hence `open` fails. Accordingly, `~open` now succeeds, and so does the query `terminal`.

The special, game-independent predicate `distinct` is computed very much like a negative subgoal.

Given: Query $L_1 \& L_2 \& \ldots \& L_n$ $(n \geq 1)$

Case: $L_1 = \texttt{distinct}(t_1, t_2)$ without variables
         • if t_1 and t_2 are syntactically different: $L_1 \& L_2 \& \ldots \& L_n \Rightarrow L_2 \& \ldots \& L_n$
         • if t_1 and t_2 are syntactically identical: $L_1 \& L_2 \& \ldots \& L_n \Rightarrow$ failure

It is crucial that only negative subgoals without variables can be subjected to the negation-by-failure computation. The same holds for the `distinct` relation. In some cases it may be necessary to reorder the subgoals to ensure that all variables have been replaced by non-variable terms before a negative literal gets selected. The GDL description of Multiple Buttons And Light (see Chapter 11) is a point in case. Consider the following encoding of a given position and a move in that game, along with one of the update rules.

```
1   true(p(1))
2   true(p(2))
3   true(p(3))
4   does(white,a(2))
5
6   next(p(X)) :- ~does(white,a(X)) & true(p(X))
```

Naturally, we expect to be able to derive two answers to the query `next(p(Z))`, namely $\{Z/1\}$ and $\{Z/3\}$, since the first and third light are not affected by the action. So let's consider the situation after the first step in a derivation.

$$
\begin{array}{lll}
& \texttt{next(p(Z))} & \text{rule 6 with } \{X/Z\} \\
\Rightarrow & \texttt{\~does(white,a(Z)) \& true(p(Z))} &
\end{array}
$$

But now if we were to ignore the restriction about negative literals with variables and chose the subgoals in the given order, then the query would fail. For the positive part of the negative literal, `does(white,a(Z))`, obviously has a successful derivation:

$$
\begin{array}{lll}
& \texttt{does(white,a(Z))} & \text{rule 4 with } \{Z/2\} \\
\Rightarrow & \square &
\end{array}
$$

If, however, we reorder the subgoals so as to ensure that before we select a negated subgoal, all its variables have been substituted, then the two answers are obtained as expected.

$$
\begin{array}{lll}
& \texttt{next(p(Z))} & \text{rule 6 with } \{X/Z\} \\
\Rightarrow & \texttt{true(p(Z)) \& \~does(white,a(Z))} & \text{rule 1 with } \{Z/1\} \\
\Rightarrow & \texttt{\~does(white,a(1))} & \text{query does(white,a(1)) fails} \\
\Rightarrow & \square &
\end{array}
$$

$$
\begin{array}{lll}
& \texttt{next(p(Z))} & \text{rule 6 with } \{X/Z\} \\
\Rightarrow & \texttt{true(p(Z)) \& \~does(white,a(Z))} & \text{rule 3 with } \{Z/3\} \\
\Rightarrow & \texttt{\~does(white,a(3))} & \text{query does(white,a(3)) fails} \\
\Rightarrow & \square &
\end{array}
$$

Fortunately, the syntactic restrictions on valid GDLs guarantee that subgoals can always be reordered in this manner. Specifically, every variable in a rule must occur in a positive literal in the body, not counting the special game-independent predicate `distinct`. Hence, you can always compute values for all variables in a clause before selecting a negative subgoal or an instance of `distinct`.

The negation-by-failure principle is recursively applied in case of nested negation, that is, when negative sub-goals arise during an attempted derivation for another negated literal. An illustrative example is given by the following excerpt from a fictitious game description.

```
1  role(red)
2  role(blue)
3  role(green)
4
5  true(free(blue))
6
7  trapped(R) :- role(R) & ~true(free(R))
8  goal(W,100) :- role(W) & ~trapped(W)
```

Only player blue can be shown to have achieved the goal in the current state, as per the nested derivation below.

	goal(blue,100)	rule 8 with {W/blue}
⇒	role(blue) & ~trapped(blue)	rule 2
⇒	~trapped(blue)	sub-derivation †
⇒	□	

---

†		
	trapped(blue)	rule 7 with {R/blue}
⇒	role(blue) & ~true(free(blue))	rule 2
⇒	~true(free(blue))	sub-derivation ‡
⇒	failure	

---

‡		
	true(free(blue))	rule 5
⇒	□	

The attempt to derive goal(red,100) fails.

	goal(red,100)	rule 8 with {W/red}
⇒	role(red) & ~trapped(red)	rule 1
⇒	~trapped(red)	sub-derivation †
⇒	failure	

---

†

	trapped(red)	rule 7 with {R/red}
⇒	role(red) & ~true(free(red))	rule 1
⇒	~true(free(red))	sub-derivation ‡
⇒	□	

---

‡

	true(free(red))	
⇒	failure	

The same is true for goal(green,100).

EXERCISES

13.1. The goal of this exercise is to unlock the mystery behind the single-layer game shown below—and then solve it.

```
role(you)

succ(1,2)
succ(2,3)
succ(3,4)
succ(4,5)
succ(5,6)
succ(6,7)
succ(7,8)

is_row(1)
is_row(Y) :-
  succ(X,Y)

base(step(1))
base(step(2))
```

```
base(step(3))
base(step(4))
base(step(5))
base(row(X,empty)) :-
  is_row(X)
base(row(X,one_coin)) :-
  is_row(X)
base(row(X,two_coins)) :-
  is_row(X)
input(you,jump(X,Y)) :-
  is_row(X) &
  is_row(Y)

init(step(1))
init(row(X,one_coin)) :-
  is_row(X)

babbage(X,Y) :-
  succ(X,Y)

babbage(X,Y) :-
  succ(X,Z) &
  true(row(Z,empty)) &
  babbage(Z,Y)

lovelace(X,Y) :-
  succ(X,Z) &
  true(row(Z,empty)) &
  lovelace(Z,Y)

lovelace(X,Y) :-
  succ(X,Z) &
  true(row(Z,one_coin)) &
  babbage(Z,Y)

turing(X,Y) :-
  succ(X,Z) &
  true(row(Z,empty)) &
  turing(Z,Y)
```

```
turing(X,Y) :-
  succ(X,Z) &
  true(row(Z,one_coin)) &
  lovelace(Z,Y)

turing(X,Y) :-
  succ(X,Z) &
  true(row(Z,two_coins)) &
  babbage(Z,Y)

legal(you,jump(X,Y)) :-
  true(row(X,one_coin)) &
  true(row(Y,one_coin)) &
  turing(X,Y)

legal(you,jump(X,Y)) :-
  true(row(X,one_coin)) &
  true(row(Y,one_coin)) &
  turing(Y,X)

next(row(X,empty)) :-
  does(you,jump(X,Y))

next(row(Y,two_coins)) :-
  does(you,jump(X,Y))

next(row(X,C)) :-
  true(row(X,C)) &
  does(you,jump(Y,Z)) &
  distinct(X,Y) &
  distinct(X,Z)

next(step(Y)) :-
  true(step(X)) &
  succ(X,Y)

terminal :-
  ~open
```

```
open :-
  legal(you,M)

goal(you,100) :-
  true(step(5))

goal(you,0) :-
  ~true(step(5))
```

(a) Answer Substitutions

Compute all successful derivations and their answer substitutions to the query init(F).

Hint: You should obtain nine different answers, which together form the initial game state. Can you draw a simple diagram to visualize it?

(b) Derivations

The key to unlocking the mystery is to understand the meaning of the three recursive relations, babbage, lovelace, and turing. To get the idea, suppose given some facts shown below.

```
true(row(1,one_coin))
true(row(2,one_coin))
true(row(3,empty))
true(row(4,empty))
true(row(5,one_coin))
true(row(6,one_coin))
true(row(7,two_coins))
true(row(8,one_coin))
```

Which of the following queries have a successful derivation?

- babbage(2,5)
- babbage(2,6)
- lovelace(1,5)
- lovelace(1,6)
- turing(1,6)
- turing(1,7)
- turing(6,8)

Can you now describe in words the meaning of turing(X,Y)?

(c) **Derivations: Legal Moves**

Next, have a look at the definition for legal(you, jump(X,Y)). In words, what are the preconditions for jumping from row X to row Y? How many actions are possible in the initial game state?

(d) **Derivations: State Update**

Now pick any one of the legal actions jump(m,n) in the initial state and compute the new state after does(you, jump(m,n)). Can you describe in words what is the effect of jumping from X to Y?

(e) **Playing**

The definitions for terminal and goal(you, 100), respectively, imply that the game ends when you are stuck (i.e., there are no more legal moves) and that you win the game when you can make the maximum of four moves before getting stuck. Find a sequence of actions that solves this game!

Hint: There is more than one solution.

(f) **Playing**

Bonus challenge: We humans are often better than computers at generalising a solution. How would you solve the game if you start with 998 coins in a row and the goal is to make the maximum of 499 moves without getting stuck?

CHAPTER 14

Analyzing Games with Logic

14.1 INTRODUCTION

Analyzing a set of rules with the aim to acquire useful knowledge about a new game is arguably the biggest, and most interesting, challenge for general game-playing systems. Most knowledge is only implicit in the description of a game and therefore needs to be learned, structured, and verified before it can be used to improve play.

The range of useful knowledge extends from basic properties, such as whether a game is zero-sum or cooperative, to expertise that can fill entire databases, such as the world's chess knowledge accumulated over centuries of play.

The uses of knowledge in a general game player are equally wide. Simple properties help to decide on the right search method, like minimax with alpha-beta pruning in case a game has been identified as zero sum with alternating moves. Structural knowledge, e.g., of symmetries in a game, can be used to accelerate any type of search. Knowledge of the value of different pieces or of different board regions can form the basis for evaluation functions, etcetera.

While the possibilities to acquire and use knowledge in a general game player are nearly limitless, in this and the following two chapters we will consider approaches that are both relatively easy to implement and at the same time (almost) universally applicable.

We begin with a solution to a very basic problem that you need to solve if, for example, you want to transform a GDL description into a more efficient representation like a propositional network. To do so you need to determine the possible values for the arguments of each function and relation in a given game description. For some relations, like `true`, `next`, and `does`, this is easily computed from the `base` and `action` relation. But for auxiliary predicates and functions, their input values need to be explicitly computed.

14.2 COMPUTING DOMAINS

Computing the domains is actually fairly easy. We just need to examine the dependencies among the arguments and variables in each game rule. This can be achieved through the construction of the *domain graph* for a given GDL description.

The vertices of this graph include all function symbols and constants that occur in the rules. In our GDL-description for Tic-Tac-Toe in Chapter 2, for example, we find the following constants and functions, listed in the order of appearance.

```
white, black, cell, x, o, b, control, mark, noop, 1, 2, 3
```

The vertices also includes one node for each argument position of each predicate and function. An example are the two nodes `row[1]` and `row[2]` for the auxiliary binary function `row` from our Tic-Tac-Toe description.

The edges in the domain graph are directed. They indicate the dependencies between the constants, functions, and argument positions.

Definition 14.1 The *domain graph* for a set of GDL rules G is the smallest directed graph (V, E) with vertices V and edges E that satisfies all of the following.

1. $c \in V$ for each constant c occurring in G.

2. $p, p[1], \ldots, p[n] \in V$ for each n-ary predicate or function symbol p occurring in G.

3. $f \rightarrow p[i] \in E$ for each occurrence of a function (or constant) f in the i-th argument of an expression p in the head of a rule in G.

4. $p[j] \rightarrow q[i] \in E$ whenever a variable X in a clause in G is shared by

   - the i-th argument of expression q in the head and
   - the j-th argument of expression p in the body.

5. E includes the three edges

   - `base[1]` $\rightarrow$ `true[1]`
   - `input[1]` $\rightarrow$ `does[1]`
   - `input[2]` $\rightarrow$ `does[2]`

As an example, recall one of the rules from our Tic-Tac-Toe description in Chapter 2.

$$\texttt{base(cell(M,N,x)) :- index(M) \& index(N)}$$

This rule gives rise to four edges in the Tic-Tac-Toe domain graph according to items 3 and 4 of Definition 14.1, as shown in the diagram below.

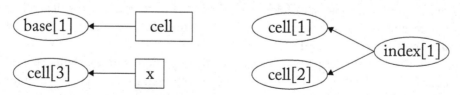

Figure 14.1: Some edges in the Tic-Tac-Toe domain graph.

The edges cell → base[1] and x → cell[3] are obtained from the head of the clause. The edges connecting index[1] to cell[1] and cell[2] follow, respectively, from the shared variables M and N.

For another example, consider the clauses defining the auxiliary concepts of a row, column, diagonal, and line.

```
line(Z) :- row(M,Z)
line(Z) :- column(N,Z)
line(Z) :- diagonal(Z)

row(M,Z) :-
  true(cell(M,1,Z)) &
  true(cell(M,2,Z)) &
  true(cell(M,3,Z))

column(N,Z) :-
  true(cell(1,N,Z)) &
  true(cell(2,N,Z)) &
  true(cell(3,N,Z))

diagonal(Z) :-
  true(cell(1,1,Z)) &
  true(cell(2,2,Z)) &
  true(cell(3,3,Z))

diagonal(Z) :-
  true(cell(1,3,Z)) &
  true(cell(2,2,Z)) &
  true(cell(3,1,Z))
```

Along with the base definition for cell,

```
index(1)
index(2)
index(3)

base(cell(M,N,x)) :- index(M) & index(N)
base(cell(M,N,o)) :- index(M) & index(N)
base(cell(M,N,b)) :- index(M) & index(N)
```

these rules determine edges in the Tic-Tac-Toe domain graph as follows.

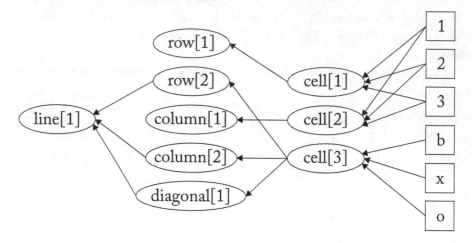

Figure 14.2: An excerpt of the Tic-Tac-Toe domain graph that determines the range of values for some of the auxiliary predicates.

The domains for the four features can easily be computed from this graph by following backwards along all possible paths from the argument positions to the constants. The possible arguments of line, say, are determined as follows.

$$
\begin{aligned}
Domain(\texttt{line[1]}) &= Domain(\texttt{row[2]}) \cup Domain(\texttt{column[2]}) \cup \\
&\quad\ Domain(\texttt{diagonal[1]}) \\
&= Domain(\texttt{cell[3]}) \\
&= \{\texttt{b},\texttt{x},\texttt{o}\}
\end{aligned}
$$

Altogether we thus obtain the domains shown in the table below.

Table 14.1: Some of the function symbols and their argument range from the rules of Tic-Tac-Toe

Function	Domain
line	$\{\texttt{b},\texttt{x},\texttt{o}\}$
row	$\{\texttt{1},\texttt{2},\texttt{3}\} \times \{\texttt{b},\texttt{x},\texttt{o}\}$
column	$\{\texttt{1},\texttt{2},\texttt{3}\} \times \{\texttt{b},\texttt{x},\texttt{o}\}$
diagonal	$\{\texttt{b},\texttt{x},\texttt{o}\}$

In the same way we can compute the domains for all other predicates and functions used in Chapter 2 to describe Tic-Tac-Toe.

Table 14.2: The argument ranges of the remaining predicates and functions from the Tic-Tac-Toe game description

Predicate/Function	Domain
role	{white,black}
index	{1,2,3}
cell	{1,2,3} × {1,2,3} × {b,x,o}
control	{white,black}
base	{cell({1,2,3}×{1,2,3}×{b,x,o}),control({white,black})}
true	{cell({1,2,3}×{1,2,3}×{b,x,o}),control({white,black})}
mark	{1,2,3} × {1,2,3}
input	{white,black} × {mark({1,2,3}×{1,2,3}),noop}
does	{white,black} × {mark({1,2,3}×{1,2,3}),noop}
init	{cell({1,2,3}×{1,2,3}×{b}),control({white})}
legal	{white,black} × {mark({1,2,3}×{1,2,3}),noop}
next	{cell({1,2,3}×{1,2,3}×{b,x,o}),control({white,black})}
goal	{white,black} × {0,50,100}

14.3 REDUCING THE DOMAINS FURTHER

While the domain graph helps to identify the range of possible values for each individual argument, it does not consider dependencies between different argument positions of a predicate or function. As a consequence, it may still generate many unnecessary instances.

Take, for example, the following rules from a GDL description of the game of Chess.

```
coordinate(a)    coordinate(b)    coordinate(c)    coordinate(d)
coordinate(e)    coordinate(f)    coordinate(g)    coordinate(h)

coordinate(1)    coordinate(2)    coordinate(3)    coordinate(4)
```

```
coordinate(5)    coordinate(6)    coordinate(7)    coordinate(8)

next_file(a,b)   next_file(b,c)   next_file(c,d)   next_file(d,e)
next_file(e,f)   next_file(f,g)   next_file(g,h)

next_rank(1,2)   next_rank(2,3)   next_rank(3,4)   next_rank(4,5)
next_rank(5,6)   next_rank(6,7)   next_rank(7,8)

adjacent(X1,X2) :-
   next_file(X1,X2)
adjacent(X1,X2) :-
   next_file(X2,X1)
adjacent(Y1,Y2) :-
   next_rank(Y1,Y2)
adjacent(Y1,Y2) :-
   next_rank(Y2,Y1)

kingmove(U,V,U,Y) :-
   adjacent(V,Y) & coordinate(U)
kingmove(U,V,X,V) :-
   adjacent(U,X) & coordinate(V)
kingmove(U,V,X,Y) :-
   adjacent(U,X) & adjacent(V,Y)
```

The rules define a king's move as going one square in either direction, that is, vertically, horizontally, or diagonally. From the domain graph we can compute the possible values for the arguments of the five predicates, as shown in Table 14.3.

But many of the instances thus obtained are unnecessary because they will never be referred to when playing the game. To begin with, the domains for both adjacent(X,Y) and kingmove(U,V,X,Y) do not distinguish between file and rank coordinates. As a consequence, the domain graph generates superfluous instances like for example adjacent(a,2) or kingmove(d,e,5,6).

Even among the instances of kingmove(U,V,X,Y) for which both (U,V) and (X,Y) are proper squares, the vast majority is not needed given the limited mobility of a king: For every (U,V) there is a maximum of eight squares (X,Y) that a king can reach in one move. If we were able to identify all combinations of arguments that do not correspond to a possible move, such as kingmove(e,2,g,7), then this would significantly reduce the number of instances that we need to compute. A simple calculation shows how much can thus be saved. There are $16^4 = 65,536$ instances of kingmove with the domain as per Table 14.2. If we respect the distinction between file and rank coordinates, this number reduces to $8^4 = 4,096$. Of these, less than $8^2 \cdot 8 = 512$

Table 14.3: The domains of some predicates from the rules of Chess as determined by the domain graph

Predicate	Domain
`coordinate`	$\{a,b,c,d,e,f,g,h,1,2,3,4,5,6,7,8\}$
`next_file`	$\{a,b,c,d,e,f,g\} \times \{b,c,d,e,f,g,h\}$
`next_rank`	$\{1,2,3,4,5,6,7\} \times \{2,3,4,5,6,7,8\}$
`adjacent`	$\{a,…,h,1,…,8\} \times \{a,…,h,1,…,8\}$
`kingmove`	$\{a,…,h,1,…,8\} \times \{a,…,h,1,…,8\} \times \{a,…,h,1,…,8\} \times \{a,…,h,1,…,8\}$

correspond to an actual king move. (The exact number is 444 because a king at the border can reach no more than five squares and only three from a corner.)

The following procedure allows you to eliminate in a given game description G most of the instances of predicates that will never be derivable.

1. Build the domain graph for G to determine the maximal range of values for all predicates and functions in the game description.

2. Let G^+ be obtained from G by

   - adding all facts `true(t)` and `does(t₁,t₂)` that follow from the domain graph; and
   - deleting all negative conditions from the rules in G.

3. For all possible predicate instances (except for the keywords `true` and `does`) with the domains obtained in step 1, check if they can actually be computed from G^+. Keep only those that can.

4. For all possible instances of `true(t)` that follow from the domain graph, keep only those for which `init(t)` or `next(t)` can be computed from G^+.

5. For all possible instances of `does(t₁,t₂)` that follow from the domain graph, keep only those for which `legal(t₁,t₂)` can be computed from G^+.

Let's see how steps 1–3 of this process will indeed eliminate all unnecessary combinations of values from Table 14.2. To begin with, out of the 49 possible instances of `next_file(X,Y)` only the seven that are given as facts can be computed from the given game rules. The same is true for `next_rank(X,Y)`. For `adjacent(X,Y)`, we can compute 28 instances (out of a total of $16 \cdot 16 = 256$), which rules out instances like, say, `adjacent(a,2)` and `adjacent(8,2)`. Finally, the only computable instances of `kingmove(U,V,X,Y)` are those 444 that correspond to actual moves by a king.

The reason for augmenting G in step 2 by all possible state propositions and all actions is that many predicates directly or indirectly depend on them. An example from Tic-Tac-Toe is shown below.

```
line(Z) :- row(M,Z)

row(M,Z) :-
  true(cell(M,1,Z)) &
  true(cell(M,2,Z)) &
  true(cell(M,3,Z))
```

With all possible instances of `true(cell(M,N,Z))` in Tic-Tac-Toe added, it follows that each of the nine combinations of arguments for `row(M,Z)` according to Table 14.1 may indeed at some point be true. The same holds for the three instances of `line(Z)`.

The reason for deleting all negative conditions in step 2 is that it would be incorrect to uphold them after having added all possible state propositions and actions. This can be seen, for example, with the rules for a draw in Tic-Tac-Toe.

```
goal(white,50) :- ~line(x) & ~line(o)
goal(black,50) :- ~line(x) & ~line(o)
```

Obviously, both `line(x)` and `line(o)` will be computable given all possible instances of `true(cell(M,N,X))`. Hence, if the negative conditions in the rules for `goal(white,50)` and `goal(black,50)` were not ignored, then we would wrongly conclude that neither of the two predicate instances will ever be derivable.

Step 4 of the procedure above is used to identify propositions that can never be true in a reachable game state. Similarly, step 5 is used to identify actions that will never be possible in a reachable state. As an example, consider a further rule from the GDL description of Chess, where the general concept of a king move from before is used to define the conditions under which a player can legally move this piece.

```
piece_owner_type(wk,white,king)
piece_owner_type(bk,black,king)

legal(P,move(K,U,V,X,Y)) :-
  true(control(P)) &
  piece_owner_type(K,P,king) &
  true(cell(U,V,K)) &
  kingmove(U,V,X,Y) &
  occupied_by_opponent_or_blank(X,Y,P) &
  ~threatened(P,X,Y)
```

Recall that we were able in step 3 to restrict the possible instances of `kingmove(U,V,X,Y)` to those for which `(X,Y)` is one square away from `(U,V)`. The very same restriction follows for `legal(white,move(wk,U,V,X,Y))` and `legal(black,move(bk,U,V,X,Y))` according to the rule just given. Hence, step 5 eliminates all instances of moves of the form `does(white,move(wk,U,V,X,Y))` and `does(black,move(bk,U,V,X,Y))` for which the two squares are not adjacent.

If steps 4 and 5 lead to a reduction in the set of state propositions and actions, then step 3 can be repeated with this reduced set in order to possibly further constrain the derivable predicate instances. This, in turn, may lead to more reductions in steps 4 and 5, so that the whole process can be iterated until no more ground predicates, state propositions, or moves are eliminated.

14.4 INSTANTIATING RULES

Once you have computed the domains of all predicates and functions, you can generate all relevant ground instantiations of the game rules, for example in order to construct a propnet. To instantiate a rule, all variables need to be substituted by appropriate values, i.e., members of the domain associated with the argument position in which each variable occurs. Variables with multiple occurrences in a rule can only be instantiated with an element from the intersection of all corresponding domains.

During the instantiation process, you can evaluate each condition of the form `distinct(X,Y)` in the body of a rule as soon as both arguments have received a value. If true, the condition itself can be removed, and if false, the entire instance of the rule should be deleted.

As an illustrative example, let's look at the Tic-Tac-Toe rule shown below.

```
next(cell(M,N,b)) :-
   does(W,mark(J,K)) &
   true(cell(M,N,b)) &
   distinct(M,J)
```

From the domain computation we know that $M,N,J,K \in \{1,2,3\}$ and $W \in \{$white,black$\}$. Hence, the rule can be instantiated in $3^4 \cdot 2 = 162$ different ways. But every third of these instances violates the condition `distinct(M,J)`, so that in fact only 108 need to be generated.

A fully instantiated game description can be reduced further in size by identifying supporting concepts that are never used by any other clause. Such instances can safely be removed together with their defining clauses. Again, this is a process that can be repeated until no further reduction is possible.

A point in case are the Tic-Tac-Toe rules defining a line.

```
line(Z) :- row(M,Z)
line(Z) :- column(N,Z)
line(Z) :- diagonal(Z)
```

```
row(M,Z) :-
  true(cell(M,1,Z)) &
  true(cell(M,2,Z)) &
  true(cell(M,3,Z))

column(N,Z) :-
  true(cell(1,N,Z)) &
  true(cell(2,N,Z)) &
  true(cell(3,N,Z))

diagonal(Z) :-
  true(cell(1,1,Z)) &
  true(cell(2,2,Z)) &
  true(cell(3,3,Z))

diagonal(Z) :-
  true(cell(1,3,Z)) &
  true(cell(2,2,Z)) &
  true(cell(3,1,Z))
```

From Table 14.1 we know that $X \in \{b,x,o\}$ for line(X). Indeed, line(b) is derivable in many reachable states, including the initial one. But the supporting concept of a line is needed only for the goal rules shown below, which do not refer to blank lines.

```
goal(white,100) :- line(x) & ~line(o)
goal(white,50) :- ~line(x) & ~line(o)
goal(white,0) :- ~line(x) & line(o)

goal(black,100) :- ~line(x) & line(o)
goal(black,50) :- ~line(x) & ~line(o)
goal(black,0) :- line(x) & ~line(o)
```

Consequently, we can delete each rule for line(X) that has been instantiated with X=b. This eliminates 7 of the 21 clauses with predicate line in the head obtained from the domains of Table 14.1. Moreover, once these have been removed there are no rules that use any of the conditions row(1,b), row(2,b), row(3,b), column(1,b), column(2,b), column(3,b), or diagonal(b). Hence, these and their defining clauses can be eliminated too.

The process of instantiating logic program rules is also known as *grounding*. Some of the techniques we described and others have been implemented in efficient systems that are commonly referred to as grounders and are not specific to GDL. An example is the grounder Gringo.

14.5   ANALYZING THE STRUCTURE OF GDL RULES

Analyzing the structure of GDL clauses has the goal to better understand the meaning of a rule by abstracting from the syntax details. This can be useful for many purposes. For instance, it enables the comparison of different formalizations of essentially the same rule. A general game player may thus be able to recognize a known game that just comes in a new guise.

Focusing on the structure of GDL rules also allows a general game-playing system to recognize symmetries in arbitrary games. As an example, the figure below illustrates two standard symmetries on the Tic-Tac-Toe board that you will be able to identify with the help of the structural rule analysis detailed below.

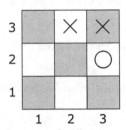 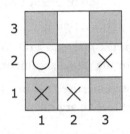 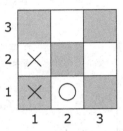

Figure 14.3: Three symmetric positions in Tic-Tac-Toe. A rotation by 180° transforms the position shown on the left-hand side into the center board. Mirroring the latter along the first diagonal results in the position on the right-hand side.

Determining symmetries like these requires to look at the game description as a whole and to see if some of its elements can be systematically exchanged with each other without affecting the meaning of any of the rules. The rules for winning or losing a game must be included in this analysis as symmetries can be broken by an asymmetric goal definition. If, say, a Tic-Tac-Toe player wins by filling a row but not a column with his or her markers, then the mirror symmetry in Fig. 14.5 would no longer apply. (The 180° rotational symmetry, in contrast, would still hold.)

14.6   RULE GRAPHS

Much like the domain computation in Section 14.2, the structural analysis can be performed on a graph constructed from the GDL rules. Specifically, the so-called *rule graph* for GDL game description is obtained through the four steps described below. Prior to applying the following definition, all variables in a game description should be renamed so that no two rules share the same variables. Constants are treated like function symbols with zero arguments. The definition is rather involved, but the example that immediately follows illustrates in detail how to construct the graph step by step.

Definition 14.2   The *rule graph* for a set G of GDL rules is a *colored* directed graph (V, E, c) whose vertices V, edges E, and vertex coloring c are obtained as follows.

1. Add a vertex for each *occurrence*, in G, of a logical connective, predicate symbol, function symbol, and variable. Connect each vertex v that represents a logical connective, predicate, or function p with the vertices for the arguments of p.

2. For each vertex v thus obtained:

   - If v stands for an n-ary function or predicate symbol p that is *not* a GDL keyword,
     - add vertices labeled $p[1], \ldots, p[n]$;
     - for each such new vertex $p[i]$, add a directed edge from $p[i]$ to the vertex that in step 1 was created for the actual argument.
   - If v stands for the binary connective ":-" or a binary GDL keyword p, add a directed edge from the first to the second argument.

3. Add a vertex for each variable or symbol p that occurs in G and is not a GDL keyword. Add a directed edge from this vertex for p to

   - each *occurrence node* for p constructed in step 1, and

   - each node for $p[1], \ldots, p[n]$ constructed in step 2 if p is a function or predicate symbol with n arguments.

4. Color the vertices such that

   - each logical connective has a unique color;
   - each GDL keyword has a unique color; and
   - all other nodes are colored in one of six colors, which depends only on their type: predicate occurrence, function occurrence, variable occurrence, argument, variable symbol, or non-variable symbol.

For illustration, recall a simple rule from our Tic-Tac-Toe game description.

```
open :- true(cell(M,N,b))
```

The result of the first step in the construction of this clause's rule graph is shown below. Vertices are depicted in different shapes to indicate different types, which will help with the coloring in the end.

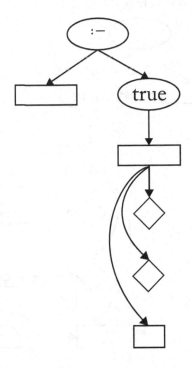

Figure 14.4: Step 1: A node for each occurrence of a logical connective (here, ":-"), predicate (open, true), function (cell, b), and variable (M, N). Directed edges lead from vertices to their arguments, if any.

In the second step, argument position nodes are added for non-keyword cell and connected to the respective occurrences. Also added is an edge between the two arguments of the logical operator ":-" (Fig. 14.5).

In step 3, nodes are added for each domain-dependent predicate symbol, function symbol, constant, and variable (Fig. 14.6).

In step 4, all vertices get colored according to their type, which completes the construction of the rule graph (Fig. 14.7). Since the *structure* of a set of rules is independent of the names given to the variables, functions, and predicates, these symbols now become irrelevant. This will enable us to compare game axiomatizations that are structurally similar and differ only in the symbols being used.

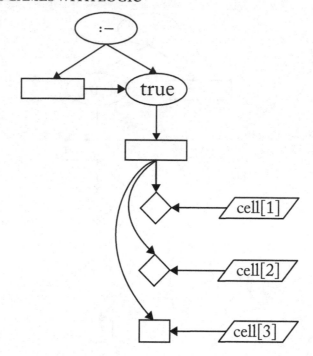

Figure 14.5: Step 2: Indicating arguments.

14.7   USING RULE GRAPHS

14.7.1   DETERMINING THE EQUIVALENCE OF GAME DESCRIPTIONS

The rule graph substitutes concrete symbols by abstract colors while maintaining the structure of the original clauses. This allows to compare syntactically different but otherwise identical game descriptions. An *isomporhism* between two colored graphs is a one-to-one mapping from the vertex set of one graph into the vertex set of the other that preserves both the edge structure and the coloring. Two graphs with an isomorphism between them are called *isomorphic*. Two GDL descriptions whose rule graphs are isomorphic describe essentially the same game.

As a simple example, the rule graph for our GDL description of Tic-Tac-Toe is isomorphic to the rule graph of any other description that just uses different coordinates, e.g., (a,a), (a,b), ...instead of (1,1), (1,2), ..., or different symbols for the two markers.

14.7.2   COMPUTING SYMMETRIES

The rule graphs can moreover be used for symmetry detection. This requires to compute *automorphisms*, that is, one-to-one mappings from a rule graph into itself that are both structure- as

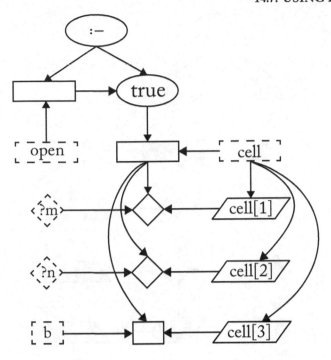

Figure 14.6: Step 3: Connecting the symbols to their occurrences.

well as color-preserving. As an example, consider exchanging two vertices of the rule graph for Tic-Tac-Toe as follows:

$$1 \to 3, 3 \to 1$$

This mapping constitutes an automorphism for the sub-graph depicted in Fig. 14.8, which is obtained from the two GDL rules shown below.

```
init(cell(1,3,b))
init(cell(3,1,b))
```

Our observation generalizes from this small sub-graph to the entire rule graph for our Tic-Tac-Toe description, which means that we have found a symmetry in this game. More specifically, we have discovered the 180° rotation symmetry from Fig. 14.5 above, which is obtained by swapping the first and third coordinate, just like in our automorphism.

The mirror symmetry along the first diagonal of the Tic-Tac-Toe board is obtained by the following exchange of two vertices in the rule graph.

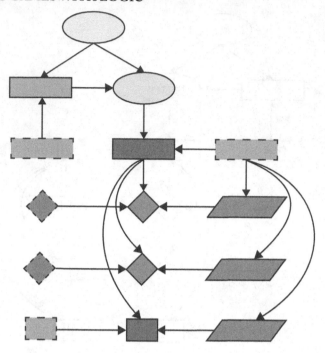

Figure 14.7: The final rule graph for open :- true(cell(M,N,b)).

$$\text{cell[1]} \rightarrow \text{cell[2]}, \text{cell[2]} \rightarrow \text{cell[1]}$$

This mapping also can be shown to be an automorphism on the graphs depicted in Figs. 14.7 and 14.9, respectively, and in fact provides an automorphism for the entire Tic-Tac-Toe rule graph.

The most common use of symmetry detection in general game players is to reduce the search space. You can, for example, prune a branch of a search tree when another branch with a symmetric joint move exists. You can also identify symmetric states and collate them in a single node in a search tree because, by definition, they must have the same value for all players.

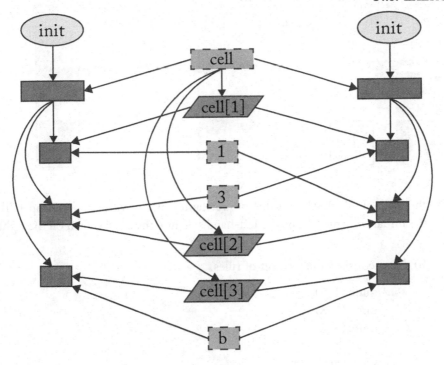

Figure 14.8: This graph remains the same when the two nodes 1 and 3 are exchanged. You can easily verify this by drawing the graph with the left-hand side and the right-hand side swapped.

14.8 EXERCISES

14.1. This exercise aims at determining the relevant ground instances of all rules of a single-player game called blocksworld (Fig. 14.9).

(a) Let's start with the rules below.

```
block(a)
block(b)
block(c)

base(table(X)) :- block(X)
base(clear(X)) :- block(X)
base(on(X,Y)) :- block(X) & block(Y)
```

From these rules draw the domain graph with nodes a,b,c,table,clear,on and block[1],base[1],table[1],clear[1],on[1],on[2].

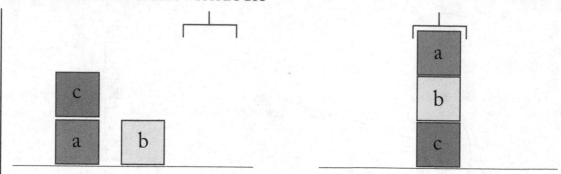

Figure 14.9: A simple single-player game of moving toy blocks with a robot gripper. The goal is to transform the initial configuration on the left-hand side into the stack shown on the right-hand side.

(b) Now consider the next set of rules.

```
succ(1,2)
succ(2,3)
succ(3,4)

base(step(1))
base(step(N)) :- succ(M,N)
```

Extend the domain graph by the nodes 1,2,3,4,step along with nodes succ[1],succ[2],step[1] and add all edges that follow from the given clauses. Use the resulting graph to determine the possible arguments for keyword base.

(c) Extend the domain graph further according to the following rules.

```
role(robot)

input(robot,stack(X,Y))   :- block(X) & block(Y)
input(robot,unstack(X,Y)) :- block(X) & block(Y)
```

Use the resulting graph to determine the possible arguments for keyword input.

(d) Complete the domain graph using the remaining rules of the game shown below.

```
init(table(a))
init(table(b))
init(on(c,a))
init(clear(b))
init(clear(c))
init(step(1))
```

```
legal(robot,stack(X,Y)) :-
  true(clear(X)) &
  true(clear(Y)) &
  distinct(X,Y)
legal(robot,unstack(X,Y)) :-
  true(clear(X)) &
  true(on(X,Y))

next(on(X,Y)) :-
  does(robot,stack(X,Y))
next(on(X,Y)) :-
  does(robot,stack(U,V)) &
  true(on(X,Y))
next(on(X,Y)) :-
  does(robot,stack(U,V)) &
  true(on(X,Y)) &
  distinct(U,X)

next(table(X)) :-
  does(robot(unstack(X,Y))
next(table(X)) :-
  does(robot(unstack(U,V)) &
  true(table(X))
next(table(X)) :-
  does(robot(stack(U,V)) &
  true(table(X)) &
  distinct(U,X)

next(clear(Y)) :-
  does(robot,unstack(X,Y))
next(clear(Y)) :-
  does(robot,unstack(U,V)) &
  true(clear(Y))
next(clear(Y)) :-
  does(robot,stack(U,V)) &
  true(clear(Y)) &
  distinct(V,Y)
```

```
next(step(N)) :-
  true(step(M)) &
  succ(M,N)

terminal :-
  true(step(4))
terminal :-
  true(on(a,b)) &
  true(on(b,c))

goal(robot,100) :-
  true(on(a,b)) &
  true(on(b,c))
goal(robot, 0) :-
  ~true(on(a,b))
goal(robot, 0) :-
  ~true(on(b,c))
```

Use the resulting graph to determine the domain of next[1]. Which element from *Domain*(base[1]) is not a member of *Domain*(next[1])?

(e) Extend the given game description G to G^+ by adding all facts true(t) and does(robot,t) that follow from the domain graph. Which of the instances of legal(robot,t) determined by the domain graph are *not* derivable from G^+ and therefore can be removed?

(f) Use all of the above to determine only the relevant instances of the rule

```
next(clear(Y)) :-
  does(robot,stack(U,V)) & true(clear(Y)) & distinct(V,Y)
```

Hint: You should obtain just 6 out of the 27 instances that without further reductions would result from domain graph.

14.2. Implement the domain graph construction, the reduction strategy, and the grounding and try it out on the Tic-Tac-Toe game description and another standard GDL game of your choice.

14.3. This exercise is concerned with finding equivalences and symmetries in a variant of the Buttons and Lights game.

(a) Draw the rule graph for the following fact.

```
role(player)
```

Why would this graph be isomorphic to the rule graph for the clause "role(white)" but not for the clause "index(1)"?

(b) Draw the rule graph for the clause below.

```
next(on(X)) :- ~true(on(X)) & does(player,toggle(X))
```

Use the rule graph method to show that this clause is structurally equivalent to the rule

```
next(p(Y)) :- does(white,q(Y)) & ~true(p(Y))
```

(c) Extend the graph for the entire game description given below.

```
role(player)

index(1)
index(2)
index(3)

base(on(X)) :- index(X)
input(player,toggle(X)) :- index(X)

legal(player,toggle(X)) :- index(X)

next(on(X)) :- ~true(on(X)) & does(player,toggle(X))
next(on(X)) :- true(on(X)) & ~does(player,toggle(X))

terminal :- true(on(1)) & true(on(2)) & true(on(3))
goal(player,100) :- true(on(1)) & true(on(2)) & true(on(3))
```

Use the rule graph method to find all symmetries in this game.

14.4. Implement the rule graph construction and try it out on the Tic-Tac-Toe game description and another standard GDL game of your choice. Search the Web for a program to compute isomorphisms of graphs and use this to determine the symmetries in each of the two games.

CHAPTER 15

Solving Single-Player Games with Logic

15.1 ANSWER SET PROGRAMMING

Single-player GDLs can be viewed as formal specifications of a logical puzzle. A perfect way to play any such game is by solving it upfront. A player can then simply unwind this solution step by step to reach a winning terminal state.

Answer Set Programming (ASP) is one of the fastest existing method for domain-independent solving of problems described in logic. It particularly lends itself to use in general game playing because the input language for ASP is very similar to GDL. Translating one into the other is therefore straightforward and can be easily automated. Moreover, a variety of ASP solvers are available for free download and can be plugged into your player.

In ASP, problems are described as collections of logic program clauses. A problem is solved by finding a minimal model for the specification.

A special type of input formula in ASP are clauses without head. Called *constraints*, they are written as

$$:- \ L_1 \& L_2 \& \ldots \& L_n$$

Their purpose is to exclude any model as solution in which all of $L_1, L_2, \ldots, L_n$ are simultaneously true.

As a first example, consider the ASP problem specification (also called *answer set program*) listed in Fig. 15.1. The clauses together have exactly one (minimal) model. To see why, observe first that p1 must obviously be true in any model according to the fact in line 1. By the constraint in line 9, p2 must be false. Hence, by clause 3, a1 must be false, because otherwise p2 would be true given that p1 is true. Knowing that a1 is false, from clause 7 it follows that b1 must be true. We thus obtain the dataset $M = \{p1, b1\}$ as the only candidate for a model. It is easy to verify that this candidate also satisfies the program rules we have not considered yet, that is, clause 6 (since ~b1 is false in M) and clause 4 (since ~p1 is false in M). Hence, we have found a minimal model of the program.

In addition to being minimal, an ASP solution must also be *supported*. It means that every atom in a candidate model needs to be the head of a clause whose body is true under the model. Our solution $M = \{p1, b1\}$ satisfies this requirement: p1 is supported by rule 1 and b1 is supported by rule 7 (since ~a1 is true in M).

```
1  p1
2
3  p2 :- a1 & p1
4  p2 :- b1 & ~p1
5
6  a1 :- ~b1
7  b1 :- ~a1
8
9     :- p2
```

Figure 15.1: An answer set program.

The example in Fig. 15.1. can be interpreted as the specification of a simple one-step puzzle. A single state proposition is true at time 1 (fact p1). Whether it is still true at time 2 (literal p2) depends on which of two actions are chosen, encoded by the literals a1 and b1. Taking the first action does not change the state proposition between times 1 and 2 (rule 3) but taking the second one does (rule 4). Clauses 6–7 together stipulate that exactly one of the two actions is chosen. Finally, the constraint in line 9 can be interpreted as the goal to make the state proposition false: answers in which p2 holds are excluded.

The model that we have just computed for this program contains b1. This provides us with a solution to the problem, namely, that the goal is achieved by taking the second action, b1.

15.2 ADDING TIME TO GDL RULES

We can generalize the idea behind the example in Fig. 15.1 to single-player games that require more than just one action. To this end we need to incorporate a linear temporal dimension into the game rules so that different instances can refer to different time points.

Let's illustrate this with a non-factored variant of the Buttons and Light game from Chapter 11. There are three lights and three buttons. Pushing the first button toggles the first light; pushing the second button moves the value of the first to the second light; and pushing the third button interchanges the second and third lights.

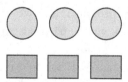

Figure 15.2: Buttons and lights.

According to the GDL description below, initially the lights are all off. The goal is to turn them on. A step counter ensures that the game terminates after six moves.

```
role(white)

base(p)
base(q)
base(r)
base(step(1))
base(step(N)) :- successor(M,N)
input(white,a)
input(white,b)
input(white,c)

legal(white,a)
legal(white,b)
legal(white,c)

init(step(1))

next(p) :- does(white,a) & ~true(p)
next(p) :- ~does(white,a) & true(p)
next(q) :- does(white,b) & true(p)
next(q) :- does(white,c) & true(r)
next(q) :- ~does(white,b) & ~does(white,c) & true(q)
next(r) :- does(white,c) & true(q)
next(r) :- ~does(white,c) & true(r)
next(step(N)) :- true(step(M)) & successor(M,N)

terminal :- true(step(7))
goal(white,100) :- true(p) & true(q) & true(r)
goal(white,  0) :- ~true(p)
goal(white,  0) :- ~true(q)
goal(white,  0) :- ~true(r)

successor(1,2)
successor(2,3)
successor(3,4)
successor(4,5)
successor(5,6)
```

```
successor(6,7)
```

Time steps can be incorporated into a set of GDL rules like these as follows.

1. Replace `init(F)` by `true(F,1)`.

2. Replace `next(F)` by `true(F,T+1)`.

3. Replace every other atom $p(...)$ by $p(...,T)$, unless

   - p is either of the keywords `role`, `base`, or `input`; or
   - p is a supporting concept that depends on neither `true` nor `does`.

4. Add the condition `time(T)` to the body of every clause to which variable T has been added.

Transforming the game rules above in this fashion results in the program clauses shown below.

```
role(white)

base(p)
base(q)
base(r)
base(step(1))
base(step(N)) :- successor(M,N)
input(white,a)
input(white,b)
input(white,c)

legal(white,a,T) :- time(T)
legal(white,b,T) :- time(T)
legal(white,c,T) :- time(T)

true(step(1),1)

true(p,T+1) :- does(white,a,T) & ~true(p,T) & time(T)
true(p,T+1) :- ~does(white,a,T) & true(p,T) & time(T)
true(q,T+1) :- does(white,b,T) & true(p,T) & time(T)
true(q,T+1) :- does(white,c,T) & true(r,T) & time(T)
true(q,T+1) :- ~does(white,b,T) & ~does(white,c,T) & true(q,T) & time(T)
true(r,T+1) :- does(white,c,T) & true(q,T) & time(T)
true(r,T+1) :- ~does(white,c,T) & true(r,T) & time(T)
true(step(N),T+1) :- true(step(M),T) & successor(M,N) & time(T)
```

```
terminal(T) :- true(step(7),T) & time(T)
goal(white,100,T) :- true(p,T) & true(q,T) & true(r,T) & time(T)
goal(white,  0,T) :- ~true(p,T) & time(T)
goal(white,  0,T) :- ~true(q,T) & time(T)
goal(white,  0,T) :- ~true(r,T) & time(T)

successor(1,2)
successor(2,3)
successor(3,4)
successor(4,5)
successor(5,6)
successor(6,7)
```

The domain for time points should be defined as ranging from 1 to some maximum solution length, say 3. Instead of explicitly adding the three facts time(1), time(2), and time(3), any standard ASP solver supports the abridged syntax

```
time(1..3)
```

Given an element T from this domain, the solver interprets the expression T+1 in the way you would expect. Note, therefore, that true is the only predicate whose time argument extends one step beyond the given horizon.

The additional time argument allows us to compute the effect of a sequence of moves within the same program. Consider, for example, the move a followed by b, then c. We can compute the sequence of game states by adding the facts shown below to our time-enriched program from above.

```
does(white,a,1)
does(white,b,2)
does(white,c,3)
```

The resulting set of clauses has exactly one minimal and supported model. This includes all of the following atoms.

Time 1	Time 2	Time 3	Time 4
legal(white,a,1)	legal(white,a,2)	legal(white,a,3)	
legal(white,b,1)	legal(white,b,2)	legal(white,b,3)	
legal(white,c,1)	legal(white,c,2)	legal(white,c,3)	
does(white,a,1)	does(white,b,2)	does(white,c,3)	
	true(p,2)	true(p,3)	true(p,4)
		true(q,3)	
			true(r,4)
true(step(1),1)	true(step(2),2)	true(step(3),3)	true(step(4),3)
goal(white,0,1)	goal(white,0,2)	goal(white,0,3)	

To understand how such models are computed, observe first that true(p,1), true(q,1), and true(r,1) must all be false in any supported model because they do not occur in the head of any clause. The same holds for does(white,b,1), does(white,c,1), and true(step(2),1), ..., true(step(7),1). It follows that true(step(1),1) and does(white,a,1)—along with the other facts in the leftmost column of the table above—are the only atoms with time argument 1 that can occur in a solution. From this we can compute all atoms with time argument 2 using the instantiated rules with the heads true(p,2), true(q,2), true(r,2), and true(step(1),2), ...true(step(7),2). The same rules but now instantiated with T = 2 can be used to generate the atoms in the solution with time argument, and so forth.

All time-independent facts are of course included in the solution too.

role(white)	base(p)	base(q)	base(r)
base(step(1))	base(step(2))	base(step(3))	base(step(4))
base(step(5))	base(step(6))	base(step(7))	input(white,a)
input(white,b))	input(white,c)	successor(1,2)	successor(2,3)
successor(3,4)	successor(4,5)	successor(5,6)	successor(6,7)

15.3 SOLVING SINGLE-PLAYER GAMES WITH ANSWER SET PROGRAMMING

We have seen how a time-expanded GDL description allows you to compute the evolution of the game state for a given sequence of actions. The goal, however, is to find some such sequence that leads to a winning terminal state. In other words, we are looking for a model in which the player takes one legal move at every time step and eventually is awarded 100 points. This is achieved with just a few more clauses.

First, we need to stipulate that the player does one action per time point until the game has terminated. In analogy to the program in Fig. 15.1, we could write the following for our example game with the three actions a, b, and c.

```
does(white,a,T) :- time(T) & ~does(white,b,T) & ~does(white,c,T)
does(white,b,T) :- time(T) & ~does(white,a,T) & ~does(white,c,T)
does(white,c,T) :- time(T) & ~does(white,a,T) & ~does(white,b,T)
```

These clauses together require that for every instance of T, exactly one element from the set {does(white,a,T), does(white,b,T), does(white,c,T)} is true in a model. Fortunately, most ASP systems support a more convenient way of specifying exactly this.

```
1 { does(white,a,T), does(white,b,T), does(white,c,T) } 1 :- time(T)
```

The numbers before and after the curly brackets, respectively, indicate the minimum and maximum number of set elements that must be contained in a solution in order for the head of this clause to be true, for any instance of T. This reduces the three clauses from above to a single one. But it still requires us to enumerate all possible moves. An even more compact encoding is obtained by implicitly, rather than explicitly, specifying the elements of a set. There is a simple way to do this by referring to another GDL keyword: because our set contains all moves M that satisfy input(white,M), we can use the following to require white to make one move at every time step.

```
1 { does(white,M,T) : input(white,M) } 1 :- time(T)
```

In fact, we do not even need to mention explicitly our player's name, white, and instead refer to the keyword role. This leads to the following rule, which is now general enough to be applicable in any single-player game.

```
1 { does(R,M,T) : input(R,M) } 1 :- role(R) & time(T)
```

A valid solution to a single-player game requires that all chosen moves be legal at the time when they are performed. This is best expressed as a constraint by which any illegal moves are ruled out.

```
:- does(R,M,T) & ~legal(R,M,T)
```

The last and of course most important requirement is to look for models in which the goal of the game is achieved. To this end, the first constraint below rules out any candidate set in which the game never reaches a terminal state. Aiming high, the second constraint *discards* any model in which the game reaches a terminal state where the player is *not* awarded 100 points.

```
:- 0 { terminal(T) : time(T) } 0

:- terminal(T), role(R), ~goal(R,100,T)
```

This completes the answer set program for solving single-player games. Coming back to our example game, Buttons and Lights, suppose the time horizon is set to 7, that is,

```
time(1..7)
```

The program then has a minimal and supported model that includes the following atoms.

does(white,a,1)	does(white,b,2)	does(white,c,3)
does(white,b,4)	does(white,a,5)	

Another model exists with moves as shown below.

does(white,a,1)	does(white,b,2)	does(white,c,3)
does(white,b,4)	does(white,b,5)	

In fact, the solution to this game comes in three variants, one for each available action at time 5. This last move is obviously irrelevant since the game has terminated at that time. But it is required according to the rule that the player has to do one action at every time point.

No solution can be found with a time horizon shorter than 5. A longer time horizon, on the other hand, will generate more models with redundant actions before or after the game has terminated. Normally, of course, the required solution length is unknown when you get a new game description. But you can start with setting it to 1 and then increment it until a model is found. This guarantees that the first computed solution is a shortest one.

15.4 SYSTEMS FOR ANSWER SET PROGRAMMING

Several ASP systems are available for free download. They are powerful enough to compute solutions to answer set programs for single-player games of medium size. Two such systems are Clasp and DLV.

15.5 EXERCISES

15.1. Minimal and supported models

Consider the answer set program in Fig. 15.1 but without rule 1. Show that this program has two minimal models. Show that only one of them is supported.

15.2. Answer set programming

Recall the rules that require a legal move at every time step.

```
1 { does(R,M,T) : input(R,M) } 1 :- role(R) & time(T)

:- does(R,M,T), not legal(R,M,T)
```

This requirement may be too strong as it includes one or more time points after the termination of a game. Extend this specification so as to require a (legal) move at time T only in case the game has not terminated at or before T.

Hint: Use the GDL keyword `terminal` but be aware that this atom may only be derivable at the time of termination but not at later points in time.

15.3. Solving a single-player game with logic

Recall the mysterious single-player game dealt with in Exercise 13.1. Now the goal is to solve this problem with an ASP system.

(a) Extend the given rules by time points as needed.

(b) Add general rules to search for a sequence of legal moves that achieves the goal.

(c) Choose a suitable time horizon and feed the resulting answer set program into an existing ASP solver.

Hints:

- Remember that this game terminates when the player is left without a legal move. Hence, you need a solution to Exercise 2.

- Familiarize yourself with the exact input syntax for the system of your choice, as this may be slightly different from the one used throughout this chapter. Specifically, the delimiter "." should be added to mark the end of a clause. And most likely, "&" and "~" need to be substituted by "," and "not", respectively.

15.4. An automatic solver for single-player games

Write a module that automatically translates single-player GDLs into answer set programs and that uses an existing ASP system to try to solve them.

CHAPTER 16

Discovering Heuristics with Logic

16.1 DISCOVERING HEURISTICS WITH ANSWER SET PROGRAMMING

Answer set programming, the topic of the previous chapter, has an interesting application beyond single-player games, namely, the discovery of useful properties of games written in GDL. You can use it to find latches, for example. Recall from Chapter 12 that a latch is a proposition that, once it becomes true (respectively, false), stays so for the rest of the game. To check that a state feature has this property, you first replace the given clauses for the initial state by the following ASP rule.

```
0 { true(F,1) : base(F) }
```

This fact acts as a *state generator* because it supports any base proposition to either be true or not at time 1. The missing number after the closing curly bracket indicates that a model can contain an unbounded number of elements from the set.

We also stipulate that each role chooses a legal move at time 1.

```
1 { does(R,M,1) : input(R,M) } 1 :- role(R)

:- does(R,M,1), ~legal(R,M,1)
```

Now suppose that we want to verify an arbitrary base proposition p to be a positive latch, that is, to stay true once it becomes true. This is achieved by trying to find a *counter-example*, that is, a model in which p is true in one state but not so in the next state.

```
counterexample :- true(p,1), ~true(p,2)
```

Finally, we need to filter out all models without such a counter-example as per the constraint below.

```
:- ~counterexample
```

Every model that an ASP system will generate is a counter-example to the hypothesis that p cannot revert back from true to false. In other words, if the solver produces *no* answer, then you have checked that the proposition *is* in fact a latch. This argument involves double negation but is perfectly valid.

In order to check that a proposition cannot revert back to true once it became false, you just need to replace the above by

```
counterexample :- ~true(p,1), true(p,2)
```

Note that you only need two time points for this proof technique, which is one of the reasons why it is very viable in practice.

16.2 GOAL HEURISTICS

Goal heuristics are based on the idea to evaluate intermediate positions according to their estimated distance to a goal—the closer a state is to our player's goal, the more promising it is. A good distance measure is often indicative of the quality of domain knowledge that a player has.

In general game playing, players begin with knowing nothing of their goal besides its pure logical description. But this information alone suffices to create a basic distance measure. An established AI technique known as *Fuzzy Logic* helps us to define it. The main idea is to equate distance with the *degree* to which a given state satisfies a goal formula.

As a motivating example, consider a larger, and hence more difficult, variant of the 8-Puzzle (cf. Chapter 5).

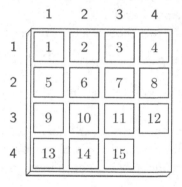

Figure 16.1: A solved 15-puzzle.

Suppose the GDL description of this game includes the goal definition depicted in Fig. 16.2. In Fuzzy Logic, the degree to which a conjunctive formula is satisfied is proportional to the number of conjuncts that are true. Hence, an intermediate position in the 15-puzzle will be judged by the number of tiles that are in the correct place. While not a perfect distance measure, it can be successfully employed as the sole parameter of the evaluation function in a depth-limited search with the effect that the player prefers moves that bring it closer to the goal configuration.

```
goal(player,100) :-
   true(cell(1,1,1)) & true(cell(2,1,2)) & true(cell(3,1,3)) &
   true(cell(4,1,4)) & true(cell(1,2,5)) & true(cell(2,2,6)) &
   true(cell(3,2,7)) & true(cell(4,2,8)) & true(cell(1,3,9)) &
   true(cell(2,3,10)) & true(cell(3,3,11)) & true(cell(4,3,12)) &
   true(cell(1,4,13)) & true(cell(2,4,14)) & true(cell(3,4,15))
```

Figure 16.2: The goal formula for the 15-puzzle.

16.3   FUZZY LOGIC

To implement a Fuzzy Logic-based goal heuristics, we first need to fix a truth value τ that satisfies $0.5 < \tau < 1$. When evaluating a GDL formula against a given state S, value τ will be assigned to all atoms $\texttt{true}(p)$ for which p holds in S. Conversely, $1 - \tau$ will be assigned to any atom $\texttt{true}(p)$ whose argument p does not hold in S.

Consider, for example, a random Tic-Tac-Toe position.

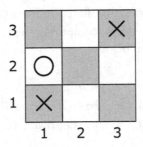

Figure 16.3: A state in Tic-Tac-Toe.

This state determines the truth values for three selected features shown in the table below, where τ has been set to 0.9.

Atom	Value
`true(cell(1,1,x))`	0.9
`true(cell(2,2,x))`	0.1
`true(cell(3,3,x))`	0.9

For the evaluation of compound formulas in Fuzzy Logic, we need to also decide on a so-called *t-norm*. This is a function $T : [0, 1] \times [0, 1] \to [0, 1]$ that is used to compute the truth value of a conjunction. (T-norm stands for: triangular norm. As such the function must satisfy $T(x, z) > T(y, z)$ whenever $x > y$ and $z > 0$.)

A common and simple t-norm is given by the product of the two arguments. For example, by multiplying the individual truth values taken from the table above we can determine the truth value of the conjunction in the rule shown below.

```
diagonal(x) :-
    true(cell(1,1,x))  &
    true(cell(2,2,x))  &
    true(cell(3,3,x))
```

The resulting value is 0.081. This can be interpreted as the degree to which the body of the clause is satisfied in the state of Fig. 16.3.

EVALUATING COMPLEX FORMULAS

Any t-norm is extensible to an evaluation function for arbitrary formulas with negation (~) and disjunction (|). For our example t-norm, computing the truth value of a compound formula follows the recursive definition shown below.

$$
\begin{aligned}
truth(\sim\!A) &= 1 - truth(A) \\
truth(A\&B) &= truth(A) \cdot truth(B) \\
truth(A\,|\,B) &= 1 - (1 - truth(A)) \cdot (1 - truth(B))
\end{aligned}
$$

The function for the disjunction can be used to compute the truth value of an atom that is defined by more than one rule. Consider, for example, the two rules defining a diagonal in Tic-Tac-Toe.

```
diagonal(x) :-
    true(cell(1,1,x)) &
    true(cell(2,2,x)) &
    true(cell(3,3,x))

diagonal(x) :-
    true(cell(1,3,x)) &
    true(cell(2,2,x)) &
    true(cell(3,1,x))
```

The body of the second clause evaluates to $0.1^3 = 0.001$ in the position shown in Fig. 16.3. Together with the value for the body of the first clause, 0.081, we obtain $0.081 + 0.001 - (0.081 \cdot 0.001) = 0.081919$ for the truth value of diagonal(x).

Analogously, we can use the rules for row(M,x) and column(N,x) to compute the truth values for each of their instances, that is, where $M, N \in \{1, 2, 3\}$. The resulting values can then be combined into the overall truth value for line(x). Again, this means to compute a disjunction from all instances of all rules for this predicate,

```
line(x)  :- row(1,x)
line(x)  :- row(2,x)
line(x)  :- row(3,x)
line(x)  :- column(1,x)
line(x)  :- column(2,x)
line(x)  :- column(3,x)
line(x)  :- diagonal(x)
```

For our example position from Fig. 16.3 we thus obtain a truth value of 0.116296 for line(x). In a similar fashion, we can compute the truth value for line(o) for the same position as 0.023797. According to the rule,

```
goal(white,100)  :- line(x) & ~line(o);
```

we can now compute the degree to which goal(white,100) is satisfied:

$$truth(\texttt{goal(white,100)}) = truth(\texttt{line(x)}) \cdot (1 - truth(\texttt{line(o)})) = 0.113529.$$

Finally, for games with positive goal values below 100, we can compute a weighted average over all goal formulas and use the Fuzzy Logic function for disjunction to obtain a single number as the overall heuristic value of a state. In Tic-Tac-Toe, where the two positive goal values are 50 and 100, respectively, we compute the weighted average for white thus:

$$100 \cdot [\, truth(\texttt{goal(white,100)}) \cdot 1$$
$$+ \; truth(\texttt{goal(white,50)}) \cdot 0.5$$
$$- \; truth(\texttt{goal(white,100)}) \cdot 1 \cdot truth(\texttt{goal(white,50)}) \cdot 0.5 \,].$$

With $truth(\texttt{goal(white,50)}) = (1 - truth(\texttt{line(x)})) \cdot (1 - truth(\texttt{line(o)})) = 0.862674$ according to the rule

```
goal(white,50)  :- ~line(x) & ~line(o);
```

this altogether results in a goal heuristics value of 49.589684 for the position shown in Fig. 16.3.

Computing the fuzzy truth values for auxiliary atoms, such as `line(x)` or `goal(x,100)`, requires to generate all relevant instances of the rules that define these atoms. This can be easily accomplished with the help of the domain analysis described in Chapter 14. The domain graph for Tic-Tac-Toe depicted in Fig. 14.2, for example, tells us which instances of the rules for `line(Z)` shown below need to be considered when computing the fuzzy truth value of `line(x)` in any given situation:

```
line(Z) :- row(M,Z)
line(Z) :- column(N,Z)
line(Z) :- diagonal(Z)
```

16.4   USING THE GOAL HEURISTICS

The primary use of the Fuzzy Logic goal heuristics is to evaluate leaf nodes in a game tree search. To see the heuristics in action, we can apply it to all successor states of the initial position in Tic-Tac-Toe to see which of the possibilities for placing the first mark looks most promising.

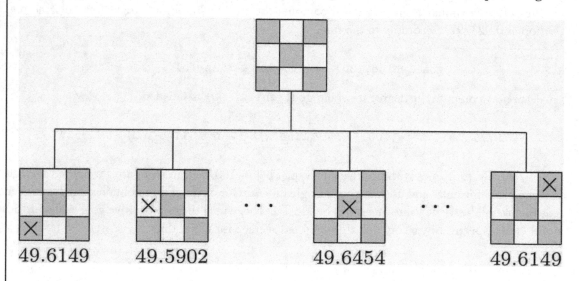

Figure 16.4: Using the goal heuristics to decide on the opening move in Tic-Tac-Toe.

Without any search beyond the first ply, it follows that the center square is the best move according to the goal heuristics. Moreover, a cell in the corner is deemed more valuable than a non-corner border cell.

16.5   OPTIMIZATIONS AND LIMITATIONS

When computing the fuzzy truth value of a defined predicate, atoms in the body of a rule can be treated differently if their truth is independent of the current state. Examples include instances of the keywords `role` and `distinct` as well as every auxiliary predicate whose definition does not depend on keyword `true`. Any such state-independent atom can be assigned truth value 1 (if it is true) or 0 (if it is false) rather than τ or $1 - \tau$. So doing simplifies the computation and also leads to sharper distinctions between different positions.

The strict application of our example t-norm for the goal heuristics can have the practical disadvantage of approaching 0 for large conjunctions even when each conjunct is true. The goal predicate in the 15-puzzle shown in Fig. 16.2, for instance, still has a low degree of $\tau^{15} = 0.9^{15} = 0.20589$ after the final goal position has been reached. In practice, it is therefore useful to introduce a *threshold* θ, which should satisfy $0.5 < \theta \leq \tau$. This threshold can be used to ensure that a true formula always has a truth value greater than 0.5. To do so, we just need to slightly extend the computation of the fuzzy truth values for complex formulas.

$$truth(\sim A) = 1 - truth(A)$$
$$truth(A\&B) = max\{truth(A) \cdot truth(B); \theta\} \qquad \text{if } truth(A) > 0.5, truth(B) > 0.5$$
$$truth(A\&B) = truth(A) \cdot truth(B) \qquad\qquad\qquad\qquad\qquad \text{otherwise}$$
$$truth(A\,|\,B) = 1 - (1 - truth(A)) \cdot (1 - truth(B))$$

The better a heuristics can distinguish between different states, the more useful it is for evaluation functions. The multiplication t-norm allows for little flexibility in this regard. A much wider range of t-norms is captured by the so-called *Yager family*. A t-norm from this family is obtained by

$$S(x, y) = (x^q + y^q)^1/q$$
$$T(x, y) = 1 - S(1 - x, 1 - y)$$

for some $0 \leq q \leq \infty$.

Any function S that is used to define a Yager t-norm T also serves as the corresponding computation rule for disjunctions. General game-playing systems that use these t-norms can adjust parameter q as well as the basic truth value τ depending on the complexity and structure of the goal formulas for different games.

LIMITATIONS

The Fuzzy Logic-based goal heuristics can be very useful for games whose goal, broadly speaking, is reachable through the accumulation of sub-goals. This extends well beyond obvious examples like the 15-puzzle shown above. If, for instance, the aim is to eliminate all of your opponent's pieces or to occupy all locations on a board, and if this can be achieved successively, then goal heuristics can provide a useful guidance for a depth-limited search.

A limitation of the simple goal heuristics is not to take into account how difficult it is to extend a partial solution to a complete one, or even if that is possible at all. In Tic-Tac-Toe, for example, every row, column, or diagonal with exactly two of our player's markers has the same fuzzy truth value, no matter whether the remaining third square is blank or already blocked. A possible solution could involve the concepts of latches and inhibitors (cf. Chapter 12). A proposition that is inhibited by an active latch can always be assigned truth value 0 because, by definition, it will never become true later in the game.

Simple goal heuristics moreover generally fail to provide useful information for games with specialized goals like, for example, checkmate in Chess, whose logical definition alone does not allow to determine the degree to which it is already satisfied in an intermediate position.

16.6 EXERCISES

16.1. Use answer set programming to discover the following properties in the Buttons and Lights game of Chapter 15.

- a. Verify that `step(1)` is a (negative) latch.

- b. Show that none of the propositions `p,q,r`, which represent the status of the lights, is a latch in this game.

- c. Write a module for your player that uses an ASP system to systematically check every base proposition whether it is a latch.

16.2. Consider the two possible replies shown below to white's opening move in Tic-Tac-Toe.

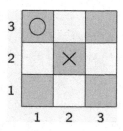

What truth value does `goal(black,100)` have in these two positions? Which move, therefore, would black take based on the goal heuristics without search?

16.3. If both players always choose their moves based on the goal heuristics without search, how would the game of Tic-Tac-Toe be played out from the following position, where black moves next?

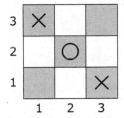

16.4. For each of the three positions shown below, what is the truth value of the formula in the body of the goal description for the 15-puzzle (cf. Figure 16.2)? Use the multiplication as t-norm to answer this question, with truth value $\tau = 0.9$ and no threshold θ.

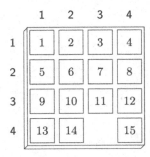

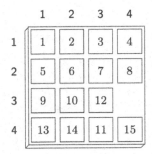

16.5. Answer the previous question for the same three positions, the same t-norm and truth value $\tau = 0.9$, but now using a threshold θ of 0.6.

16.6. Implement the Fuzzy Logic goal heuristics in your player and try it out on the 15-puzzle using minimax search.

CHAPTER 17

Games with Incomplete Information

17.1 INTRODUCTION

Our input language GDL has been designed with the assumption that the players are informed about each others' moves after every round. The effects of these moves are completely and unambiguously determined by the game rules, so that players can always compute the subsequent game state from the current one. Given that they know the initial position, they effectively have complete knowledge of the state throughout a game.

This is adequate for classical board games like Chess, Checkers, and Go. But many games of interest use moves with indeterminate effect, like rolling a die. Others are characterized by incomplete and asymmetric information. In the chess variant known as Kriegspiel, for example, the two players only can see the pieces of their own color. Most card games combine randomized moves (shuffling) with information asymmetry (you can't see cards dealt to the other players). The poker variant Texas Hold'em is a point in case.

Truly general-game playing therefore requires the extended game description language GDL-II, which enables descriptions of games with nondeterministic moves and where the players have incomplete information about the game state.

17.2 GDL-II

GDL-II stands for "GDL for games with incomplete information." By incompleteness we mean that players do not know the full game state. Mathematical game theorists draw a finer distinction between what they call imperfect-information games and those of incomplete information. We do not differentiate between the two and just mention that both can be modeled in GDL-II.

The extended game description language uses two additional keywords:

percept(r,p)	means that p is a percept of player r in the game.
sees(r,p)	means that player r perceives p in the next state.

The keyword percept(r,p) is used to define all possible percepts for a player in the same way as GDL-keyword input(r,a) describes the range of available moves. Percepts can be anything from the move by another role to a specific state feature such as the value of a card dealt face-down to the player. Even communication between players, both private and public, can be described with the help of percepts that are triggered by certain moves.

What a player actually perceives is described by the predicate sees(r,p) and generally depends on both the current state and a joint move, much like GDL-keyword next(f) does. Percepts are, by definition, private. Hence, they can be used to model games with imperfect and asymmetric information.

Regular GDL games can be easily modeled in GDL-II by adding the general game rule

$$\texttt{sees(P,move(Q,M)) :- role(P) \& does(Q,M)}$$

With this clause, all players are informed about each others' moves in every round. This information suffices to maintain complete knowledge of the position throughout the game, just as in regular GDL.

To describe games with chance moves, we also introduce a new auxiliary keyword:

random   means a pre-defined role that moves purely randomly.

By definition, the role of the random player is to always choose with uniform probability among its legal moves. Nondeterministic actions can thus be modeled as moves whose actual effect depends on the move simultaneously chosen by this game-independent role.

17.3   BLIND TIC-TAC-TOE

As an example of how to describe imperfect-information games in GDL-II, let us look at a variation of Tic-Tac-Toe where, much like in Kriegspiel, the players don't get to see each others' moves. Of course it may then happen that a player intends to mark a cell that's already been occupied. In that case, the move shall have no effect but the player will be informed about it. To make the game fairer, both players mark concurrently. If they happen to choose the same cell at the same time, then the toss of a coin determines who is successful.

To begin with, our new two-person game, which we call Blind Tic-Tac-Toe, features three roles, where the new random player is needed to simulate the coin toss.

```
role(white)
role(black)
role(random)
```

Note that you are required to declare random as a role whenever you want to use it in a game, even though it is a pre-defined one. This guarantees upward compatibility with GDL.

The initial state is similar to classical Tic-Tac-Toe (see Chapter 2) but without the state feature to indicate whose turn it is, which is no longer needed now that white and black move concurrently. A new feature, tried(P,M,N), records every attempt by a player to mark a cell. Players will only be allowed to try each cell once, which will help to ensure that the game terminates.

```
index(1)
index(2)
index(3)

base(cell(M,N,x)) :- index(M) & index(N)
base(cell(M,N,o)) :- index(M) & index(N)
base(cell(M,N,b)) :- index(M) & index(N)

base(tried(white,M,N)) :- index(M) & index(N)
base(tried(black,M,N)) :- index(M) & index(N)

init(cell(1,1,b))
init(cell(1,2,b))
init(cell(1,3,b))
init(cell(2,1,b))
init(cell(2,2,b))
init(cell(2,3,b))
init(cell(3,1,b))
init(cell(3,2,b))
init(cell(3,3,b))
```

Next, we define legality in Blind Tic-Tac-Toe. In each round, both white and black may choose any cell that they have not already tried. At the same time a coin toss is simulated, the result of which is used to break the tie in case both players attempt to mark the same cell. Accordingly, we call this random move a tiebreak.

```
input(white,mark(M,N)) :- index(M) & index(N)
input(black,mark(M,N)) :- index(M) & index(N)

input(random,tiebreak(x))
input(random,tiebreak(o))

legal(white,mark(X,Y)) :-
  index(X) &
  index(Y) &
  ~true(tried(white,X,Y))

legal(black,mark(X,Y)) :-
  index(X) &
  index(Y) &
```

```
    ~true(tried(black,X,Y))

  legal(random,tiebreak(x))
  legal(random,tiebreak(o))
```

Next, we look at the new update rules for the game. Any try to mark a cell is recorded. A cell is marked with an "x" or an "o" if the corresponding player chooses that cell and if the cell is blank and not simultaneously targeted by the other player. If both players aim at the same cell, then that cell ends up being marked according to the result of the random tie-breaking move. Finally, if a cell contains a mark, then it retains that mark on the subsequent state; and if a cell is blank and neither player attempts to mark it, then it remains blank.

```
  ; any new attempt to mark a cell is recorded
  next(tried(W,M,N)) :-
    does(W,mark(M,N))

  ; all recorded attempts are remembered
  next(tried(W,M,N)) :-
    true(tried(W,M,N))

  ; white is successful in marking a blank cell
  ; when black moves in a different column
  next(cell(M,N,x)) :-
    does(white,mark(M,N)) &
    true(cell(M,N,b)) &
    does(black,mark(J,K)) &
    distinct(M,J)

  ; white is successful in marking a blank cell
  ; when black moves in a different row
  next(cell(M,N,x)) :-
    does(white,mark(M,N)) &
    true(cell(M,N,b)) &
    does(black,mark(J,K)) &
    distinct(N,K)

  ; black is successful in marking a blank cell
  ; when white moves in a different column
  next(cell(M,N,o)) :-
    does(black,mark(M,N)) &
```

```
    true(cell(M,N,b)) &
    does(white,mark(J,K)) &
    distinct(M,J)

; black is successful in marking a blank cell
; when white moves in a different row
next(cell(M,N,o)) :-
    does(black,mark(M,N)) &
    true(cell(M,N,b)) &
    does(white,mark(J,K)) &
    distinct(N,K)

; if both players aim at the same cell, then that cell
; gets marked by the result of the random tiebreak move
next(cell(M,N,W)) :-
    true(cell(M,N,b)) &
    does(white,mark(M,N)) &
    does(black,mark(M,N)) &
    does(random,tiebreak(W))

; markings are forever
next(cell(M,N,x)) :-
    true(cell(M,N,x))

next(cell(M,N,o)) :-
    true(cell(M,N,o))

; a cell remains blank if no player attempts to mark it
next(cell(M,N,b)) :-
    true(cell(M,N,b)) &
    ~marked(M,N)

marked(M,N) :-
    does(W,mark(M,N))
```

GDL-II is based on the assumption that players are no longer automatically informed about any other players' moves. Thus, without additional hints our Blind Tic-Tac-Toe players would be completely oblivious as to whether any of their attempts to mark a cell was successful. According to the following rules, which define the players' percepts, they are provided with exactly that but no more information.

```
percept(white,ok)
percept(black,ok)

; players get ok when they mark a blank cell
; in a different column from where their opponent moves
sees(R,ok) :-
  does(R,mark(M,N)) &
  true(cell(M,N,b)) &
  does(S,mark(J,K)) &
  distinct(M,J)

; players get ok when they mark a blank cell
; in a different row from where their opponent moves
sees(R,ok) :-
  does(R,mark(M,N)) &
  true(cell(M,N,b)) &
  does(S,mark(J,K)) &
  distinct(N,K)

; white gets ok when he marks a blank cell
; and the random tiebreak went to his side
sees(white,ok) :-
  does(white,mark(M,N)) &
  true(cell(M,N,b)) &
  does(random,tiebreak(x))

; black gets ok when he marks a blank cell
; and random tiebreak went to his side
sees(black,ok) :-
  does(black,mark(M,N)) &
  true(cell(M,N,b)) &
  does(random,tiebreak(o))
```

By these rules a player sees "ok" after attempting to mark a cell that was indeed blank and that was not simultaneously targeted by the opponent. The very same will be seen by the player in whose favor the tie was broken, provided again that the cell being aimed at was empty. Since the percepts are identical in both cases, players cannot distinguish between them. Hence, they will not know, for example, if their opponent attempted (and failed) to mark the same cell at the same time. Even when both cases apply together, the percept will not change.

Still, the players obtain enough information to infer which of the board's cells carry their own marks. In turn, this allows them to determine exactly their legal moves according to the rules from above. Generally speaking, a good GDL-II game description should always provide players with sufficient information for them to know their legal moves and also to know whether a game has terminated and what their result is.

In our example game, the absence of the percept "ok" tells players that they were unsuccessful in their attempt to mark a cell. A smart player can even conclude that the cell in question must now be occupied by their opponent, because either the cell already was marked before, or the tie break decided in the other player's favor.

The terminal condition for our blind version of Tic-Tac-Toe can be taken as is from the description of the standard game. Also the rules for the goals are the same as before but need to be extended to the possibility that both players complete a line in the same round.

```
terminal :- line(x)
terminal :- line(o)
terminal :- ~open

goal(white,100) :- line(x) & ~line(o)
goal(white, 50) :- line(x) & line(o)
goal(white, 50) :- ~open & ~line(x) & ~line(o)
goal(white,  0) :- ~line(x) & line(o)

goal(black,100) :- ~line(x) & line(o)
goal(black, 50) :- line(x) & line(o)
goal(black, 50) :- ~open & ~line(x) & ~line(o)
goal(black,  0) :- line(x) & ~line(o)
```

The supporting concepts line(z) and open shall be defined as before; see Chapter 2.

17.4   CARD GAMES AND OTHERS

The two new keywords in GDL-II can be used to describe all kinds of card games, which are typically characterized by both randomness (shuffle) and information asymmetry (individual hands). For example, a single card dealt face down to a player can be specified thus.

```
legal(random,deal(Player,Card)) :-
    role(Player) &
    distinct(Player,random) &
    true(indeck(Card))
```

```
next(holds(Player,Card)) :-
   does(random,deal(Player,Card))
sees(Player,Card) :-
   does(random,deal(Player,Card))
```

Here, only the player who is dealt the card can see it. Multiple cards can be handed out in a single move that takes each card as a separate argument and of which players only get to see the argument position for their card.

In contrast, a card dealt face-up, say like in Texas Hold'em, would be described as follows.

```
legal(random,deal(river(C))) :-
   true(indeck(C))

next(river(C)) :-
   does(random,deal(river(C)))
sees(P,river(C)) :-
   role(P) &
   distinct(P,random) &
   does(random,deal(river(C)))
```

With the last rule all players are informed about the river card.

Games involving communication among players, both public and private, can also be described in GDL-II. For example, as part of a negotiation a player P may offer a player Q to exchange an item C for another item D. This can be formalized by the GDL-II clauses below.

```
legal(P,ask(Q,trade(C,D))) :-
   true(has(P,C)) &
   true(has(Q,D)) &
   distinct(P,Q) &
   distinct(C,D)

sees(Q,offer(P,C,D)) :-
   does(P,ask(Q,trade(C,D)))
```

Under these rules, their communication is private: only the addressee gets to see the offer.

17.5   GDL-II GAME MANAGEMENT

GDL-II requires a modified protocol for running a game so that players are no longer automatically informed about everyone's moves after each round (cf. Chapter 3). Rather, each player gets his or her individual percept, or percepts, as determined by the game rules for the new keyword sees.

Shown below is the format of the Play message for GDL-II games in the current GGP communication language. The parameters are: (1) the usual match identifier; (2) an integer counting the number of turns; (3) the move that was executed by the player in the last turn; and (4) a list of the player's percepts.

$$play(\textit{id, turn, move, percept})$$

If the player receives no information according to the game rules, the *percept* field is []. The number of turns and the confirmation of the previous move help players recover from communication and other errors, when the game master selected a move on their behalf. On the first request, *turn* is 0, *move* is nil, and the *percept* field is [].

The specification of the Stop message is likewise modified.

$$stop(\textit{id, turn, move, percept})$$

Here is a sample of messages for a quick game of Blind Tic-Tac-Toe. As always, the Game Manager initiates the match by sending a start message to all of the players with their individual role, the rules of the game, and the values for the *startclock* and *playclock*. The players then respond with ready.

Game Manager to Player x:	start(m23,white,[...],10,10)
Game Manager to Player y:	start(m23,black,[...],10,10)
Player x to Game Manager:	ready
Player y to Game Manager:	ready

The manager starts play by sending an initial play message to all of the players. In this case, the first player responds with the action mark(1,1) while the second player chooses mark(2,3).

Game Manager to Player x:	play(m23,0,nil,[])
Game Manager to Player y:	play(m23,0,nil,[])
Player x to Game Manager:	mark(1,1)
Player y to Game Manager:	mark(2,3)

The Game Manager checks that these actions are legal, chooses a legal move for random, updates the state of the game according to this joint move, and then sends play messages to the players to solicit their next actions. Since the players aimed at different cells, they both get their "ok". On the next step, the two players attempt to mark the same cell and both play mark(2,2).

Game Manager to Player x:	play(m23,1,mark(1,1),[ok])
Game Manager to Player y:	play(m23,1,mark(2,3),[ok])
Player x to Game Manager:	mark(2,2)
Player y to Game Manager:	mark(2,2)

Again, the Game Manager checks legality, randomly chooses a legal move for `random`, updates its state, and sends a `play` message requesting the players' next actions. In this case, the random move decided in favor of the first player, which therefore is the only player to again see "ok". The first player takes advantage of the situation and plays `mark(3,3)` while the second player chooses `mark(1,3)`.

Game Manager to Player x:	`play(m23,2,mark(2,2),[ok])`
Game Manager to Player y:	`play(m23,2,mark(2,2),[])`
Player x to Game Manager:	`mark(3,3)`
Player y to Game Manager:	`mark(1,3)`

With this move, the game is over. The Manager lets the players know by sending a suitable `stop` message, stores the results in its database for future reference, and terminates.

Game Manager to Player x:	`stop(m23,3,mark(3,3),[ok])`
Game Manager to Player y:	`stop(m23,3,mark(1,3),[ok])`
Player x to Game Manager:	`done`
Player y to Game Manager:	`done`

17.6   PLAYING GDL-II GAMES: HYPOTHETICAL STATES

Most of the techniques that we have dealt with in this book can be applied to GDL-II unchanged: factoring, the discovery of heuristics, and logical reasoning all operate on the elements of a game description. Their syntax and semantics is the same in both complete—and incomplete—information games.

But there are additional challenges specific to GDL-II. General game-playing systems need to be able also to draw the right conclusions from their partial observations. An example of this was mentioned above, when the absence of the observation "ok" implied that a particular cell had to be occupied by the opponent. This illustrates the kind of logical reasoning that GDL-II games require.

Players also need to evaluate the value of a move under incomplete knowledge of the state. They can do so, at least in principle, by computing a complete table of all possible positions after each round. This table is called an *information set* in mathematical game theory. It is constructed in a similar way to a game tree. Beginning with the initial position, which is fully known, a player can compute the resulting states for all combinations of legal moves. Later in the game, the player can generate every legal successor state for every possible current state. The percepts made after every round serve as filters. They allow to exclude any element from the information set that, according to the game rules, would have led to an observation different from the actual one.

Figure 17.1 shows an example of the possible states from white's perspective after the first round in Blind Tic-Tac-Toe.

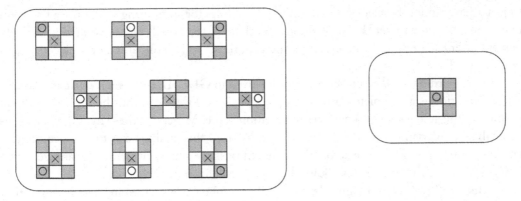

Figure 17.1: The set of positions that white—the player playing crosses—considers possible after selecting the middle square in the first move and, respectively, seeing ok (left-hand side) or seeing nothing (right-hand side).

The depicted information sets illustrate how a general game-playing system can draw logical conclusions from a complete table of possible states. Square (2,2) carries an "x" in all elements of the first set. It follows that white's move must have been successful. On the other hand, (2,2) carries an "o" in each element of the second set, which in fact is a singleton. Again, white can deduce everything that follows logically from the absence of the expected percept. Black of course will likewise know that the cell in the middle has been marked in this case. Unlike white, however, black has no way of knowing that this is the only square that is occupied after the first round.

Provided the set of possible states remains sufficiently small throughout a game, a player can use this set to determine all moves that are guaranteed to be legal. The evaluation of a move can likewise be based on the entire information set. A player can, for instance, follow the minimax principle and always choose the move with the highest minimum value across all possible positions.

Blind Tic-Tac-Toe is an example of a game with a manageable table of possible states. The possible moves by the other players, including random, lead to further branching as the game progresses. On the other hand, the players gain information after each round, which helps to contain the growth of the information set.

But this is obviously not the case for larger imperfect-information games, like Chess without revealing opponents' moves or where a deck of cards is shuffled at the beginning. Maintaining the complete information set is practically impossible in these games.

17.7 SAMPLING COMPLETE STATES

A simple solution is to apply a technique similar to Monte Carlo tree search (cf. Chapter 8). Rather than generating all possible states, a player can restrict the computation of successor states to just a few, randomly selected joint moves. This allows for computing with only a few elements from the information set. Each of these hypothetical states will be tested against the player's percept after each round. If inconsistent, the hypothesis will be discarded or replaced by another randomly selected possible state.

Every ordinary GDL player can thus be lifted to GDL-II, since each element picked from the information is a complete state. A single hypothetical state can thus be treated by the player as if it was the actual state in a perfect-information game. Moreover, these hypothetical states can be analyzed independently. The individual results from this analysis can then be combined into an expect value for each move across the different possible states. A further advantage is that the independence of the hypotheses allows the easy parallelization of this process.

But playing GDL-II games by randomly sampling complete states has its limitations. A player that considers only a small subset of all possible states may draw wrong conclusions. Even the mere legality of a move is not guaranteed by its being legal in just a few sample states. A move that seems promising in some states may likewise turn out not to be a good move if all possible positions would be considered.

A further and more fundamental disadvantage lies in the implicit assumption of complete information when playing with individual elements of the information set. In so doing, we completely ignore the difference between knowledge and the lack thereof. As a consequence, a move will never be considered useful if its sole purpose is to gain information. This is so because additional information has no value for any hypothetical state that assumes complete knowledge anyway.

Let's look at a very simple example of a single-player game that demonstrates how little the sampling technique values knowledge and how it will never choose an information-gathering move. Play commences with the random player choosing a red or blue wire to arm a bomb. The player, which shall be called agent, may then choose whether or not to ask which wire was used; asking carries a cost of 10 points to the final score. Finally, the agent must cut one of the wires to either disarm—or accidentally detonate—the bomb.

```
role(random)
role(agent)

input(R,noop) :- role(R)
input(agent,ask)
input(random,arm(C)) :- color(C)
input(agent,cut(C)) :- color(C)

color(red)
color(blue)

base(step(1))
```

```
base(step(N))   :- succ(M,N)
base(armed(C)) :- color(C)
base(score(S)) :- payoff(S)

succ(1,2)
succ(2,3)
succ(3,4)

payoff(90)
payoff(100)

init(step(1))

legal(random,arm(C)) :- true(step(1)) & color(C)
legal(random,noop)   :- ~true(step(1))
legal(agent,noop)    :- true(step(1))
legal(agent,noop)    :- true(step(2))
legal(agent,ask)     :- true(step(2))
legal(agent,cut(C))  :- true(step(3)) & color(C)

sees(agent,C) :- does(agent,ask) & true(armed(C))

next(step(N))     :- true(step(M)) & succ(M,N)
next(score(90))   :- does(agent,ask) & true(step(2))
next(score(100)) :- does(agent,noop) & true(step(2))
next(score(S))    :- true(score(S))
next(armed(C))    :- does(random,arm(C))
next(armed(C))    :- true(armed(C))
next(explodes)    :- does(agent,cut(C)) & true(armed(C))

terminal :- true(step(4))

goal(agent,S) :- true(score(S)) & ~true(explodes)
goal(agent,0) :- true(explodes)
```

Shown below (Fig. 17.2) is the full game tree, including the outcomes for the agent at each terminal node. If a player samples the two states in the information set after round 1 and assumes complete knowledge in each of them, then in either case he believes he knows which color has been used to arm the bomb. Hence, he never asks the question in this game since it carries a penalty that he thinks he can avoid (due to superficial agreement of the samples). But

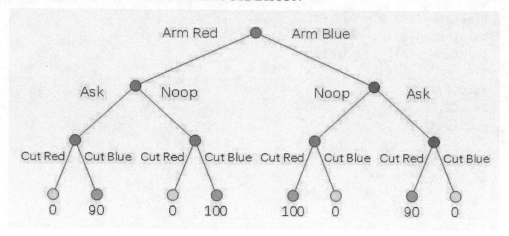

Figure 17.2: The exploding bomb game tree.

after playing noop in round 2 the two samples (that is, the two gray nodes at depth 2 in Fig. 17.2) do not agree on which move to take next, so that the player can do no better than choose randomly and thus obtain an average outcome of 50.

A variation of the Exploding Bomb game shows that a player also will never discount a move that gives away valuable information to opponents, for the same reason. In this version our player plays the arming agent—that chooses which wire arms the bomb—and also decides whether to tell his opponent which wire to cut. Freely providing this information carries a reward of 10 points.

```
role(agent)
role(opponent)

input(R,noop) :- role(R)
input(agent,tell)
input(agent,arm(C)) :- color(C)
input(opponent,cut(C)) :- color(C)

color(red)
color(blue)

base(step(1))
base(step(N))  :- succ(M,N)
base(armed(C)) :- color(C)
base(score(S)) :- payoff(S)
```

```
succ(1,2)
succ(2,3)
succ(3,4)

payoff(90)
payoff(100)

init(step(1))

legal(agent,arm(C))     :- true(step(1)) & color(C)
legal(agent,noop)       :- true(step(2))
legal(agent,tell)       :- true(step(2))
legal(agent,noop)       :- true(step(3))
legal(opponent,noop)    :- ~true(step(3))
legal(opponent,cut(C))  :- true(step(3)) & color(C)

sees(opponent,C) :- does(agent,tell) & true(armed(C))

next(step(N))       :- true(step(M)) & succ(M,N)
next(score(100))    :- does(agent,tell) & true(step(2))
next(score(90))     :- does(agent,noop) & true(step(2))
next(score(S))      :- true(score(S))
next(armed(C))      :- does(agent,arm(C))
next(armed(C))      :- true(armed(C))
next(explodes)      :- does(opponent,cut(C)) & true(armed(C))

terminal :- true(step(4))

opposite(90,10)
opposite(100,0)

goal(agent,S)     :- true(explodes) & true(score(S))
goal(agent,T)     :- ~true(explodes) & true(score(S)) & opposite(S,T)
goal(opponent,S)  :- ~true(explodes) & true(score(S))
goal(opponent,T)  :- true(explodes) & true(score(S)) & opposite(S,T)
```

If our agent samples the information set after his first move and again makes the implicit assumption that both players play with complete information, then he believes that the opponent knows the right color anyway. Hence, he always tells to avoid the penalty and thus is guaranteed to lose against a rational opponent (with a score of 10 vs. 90 points) while withholding the information would lead to a better outcome on average as the opponent has a mere 50/50 chance.

The two example games in this section are better approached by taking into account the entire information set of a player to decide on a move. But to reason correctly about one's own knowledge (and that of the other players) remains one of the many major challenges in general game playing with incomplete information for practical games where information sets are too large to be maintained explicitly.

17.8 EXERCISES

17.1. In Blind Tic-Tac-Toe, how many states are possible at the end of round 2 after you

a. played `mark(2,2)` then `mark(1,3)` and your percepts were `nil` followed by `(ok)`?
b. played `mark(2,2)` then `mark(1,3)` and your percepts were `nil` followed by `nil`?
c. played `mark(2,2)` then `mark(1,3)` and your percepts were `(ok)` followed by `nil`?

17.2. Modify the rules of Blind Tic-Tac-Toe so as to describe a version without the `random` role, where

- both players simultaneously choose a cell to mark;

- they can choose any cell that is not already occupied with their own marker;

- if they choose the same cell in the same move, then this cell remains blank;

- otherwise, a player succeeds in marking a cell only if that cell was empty;

- they see "ok" if and only if their attempt to mark a cell was successful.

17.3. Describe the following card game in GDL-II.

A deck initially has 13 cards: ♣2, ♣3, …, ♣10, ♣J, ♣Q, ♣K, ♣A. Two players, `alice` and `bob`, are randomly dealt one card each from the deck. The first player then decides whether to bet or fold.

- If she folds, the second player gets $5.

- If she bets, the second player can decide to fold, check, or raise.

  – If he folds, neither player gets anything.

  – If he checks, the player with the higher card gets $5.

  – If he raises, the first player can either fold or check.

    * If she folds, the second player gets $5.

    * If she checks, the player with the higher card wins $10.

The players get to see their opponent's card immediately after one of them has decided to check. If, however, the round ended with one of them folding, then the players are not informed about each other's hand. The game continues with two new cards dealt from the reduced deck, but now bob has to make the initial choice between betting and folding. The game ends after six rounds, when there is only one card left in the deck. The player wins who has amassed the higher total amount.

17.4. Extend the game description from Exercise 3 by giving both players $100 at the start. When betting or raising, the players can decide how much they want to put in. When they check they have to match the previous bet (or raise, respectively), or go all-in. The minimum opening bet is always $5. The game ends prematurely when one of the players went broke.

17.5. Invent a simple (but non-trivial) game with imperfect information of your own. Describe it in English. Write a GDL-II rulesheet for the game. Give a move history that takes the game from the initial state to a terminal state. Use the GDL stepper to show that your history works.

17.6. Implement the stochastic simulation of possible states as a bolt-on solution to your GDL player. Try your player out on one of the standard GDL-II games.

CHAPTER 18

Games with Historical Constraints

18.1 INTRODUCTION

One of the distinctive features of GDL is its *Markov* character—the truth values of all state-dependent conditions are defined entirely in terms of the truth values of conditions on the current state and/or the immediately preceding state. There is no explicit dependence on other states of the world.

In some games and in many real-word applications, it is more convenient to define state-dependent relations (such as legality, reward, and termination) in terms of multiple preceding states. For example, in Chess, we need to express the fact that a player may not castle if it has moved either its king or rook on any preceding step.

Now, for finite games, it is always possible to transform such conditions into Markov conditions by adding information to the state of the game. Unfortunately, this is sometimes inconvenient; and in many real-world applications, this is not possible, since the state description is often controlled by others. In such cases, we need a language to allow us to express non-Markov conditions directly, without such state modifications.

In this chapter, we examine a language called System Definition Language (or SDL), which supports this level of expressiveness. In the next section, we give the details of the language. We then show how to use it in the context of a couple of examples.

18.2 SYSTEM DEFINITION LANGUAGE

System Description Language (or SDL) is a non-Markov variant of GDL. Like GDL, descriptions take the form of open logic programs. The only difference is in the game-independent vocabulary, and even this is very similar to that of GDL.

First of all, SDL includes all of the structural relations in GDL without change.

`role(r)`	means that r is a role in the game.
`base(p)`	means that p is a base proposition in the game.
`percept(r,a)`	means that p is a percept for role r.
`input(r,a)`	means that a is an action for role r.

To these basic structural relations, we add a couple of relations for talking about steps.

step(s)	means that s is a step.
successor(s_1,s_2)	means that step s_1 comes immediately before step s_2.

The relations true, sees, does, legal, goal, and terminal are all the same as in GDL *except that* each is augmented with a step argument.

true(p,s)	means that the proposition p is true on step s.
sees(r,p,s)	means that role r sees percept p on step s.
does(r,a,s)	means that role r performs action a on step s.
legal(r,a,s)	means it is legal for role r to play action a on step s.
goal(r,n,s)	means that player has utility n for player r on step s.
terminal(s)	means that the state on step s is terminal.

And that's it; that's the entire language. We no longer need init and next. Why? The truth of propositions in the initial state can be stated using true with the first step as the step argument; and update rules, formerly written using next, can be stated using true and successor. Just how this works should become clear from the examples given below.

18.3 EXAMPLE–TIC-TAC-TOE

As an illustration of SDL, let's see how we can use the language to describe the game of Tic-Tac-Toe. As we have seen, we can describe the game adequately in GDL. The point of rewriting the description in SDL is to underscore the similarities and differences.

In SDL, as in GDL, we use the ternary function constant cell together with a row m and a column n and a mark w to designate the proposition that the cell in row m and column n contains w where w is either an x or an o or a b (for blank); and we use the unary function constant control to state that it is that role's turn to mark a cell. The binary function mark together with a row m and a column n designates the action of placing a mark in row m and column n, and the object constant noop refers to the act of doing nothing.

The first step in writing an SDL description is the same as that in GDL: we enumerate the structural components of the game—roles, base propositions, percepts (there are none in this case), and actions. In SDL, these descriptions are exactly the same as in GDL.

```
role(white)
role(black)

base(cell(X,Y,W)) :-
   index(X) &
   index(Y) &
   filler(W)
```

```
base(control(W)) :-
  role(W)

input(W,mark(X,Y)) :-
  role(W) &
  index(X) &
  index(Y)

input(W,noop) :-
  role(W)

index(1)
index(2)
index(3)

filler(x)
filler(o)
filler(b)
```

To these basic components, we add a set of steps and a suitable successor function. In this case, we use the natural numbers 1, ..., 10 with their usual successor function.

```
step(1)
step(2)
step(3)
step(4)
step(5)
step(6)
step(7)
step(8)
step(9)
step(10)

successor(1,2)
successor(2,3)
successor(3,4)
successor(4,5)
successor(5,6)
successor(6,7)
successor(7,8)
```

```
successor(8,9)
successor(9,10)
```

The initial state of the game is written using the `true` relation and the first step (in this case the natural number 1).

```
true(cell(1,1,b),1)
true(cell(1,2,b),1)
true(cell(1,3,b),1)
true(cell(2,1,b),1)
true(cell(2,2,b),1)
true(cell(2,3,b),1)
true(cell(3,1,b),1)
true(cell(3,2,b),1)
true(cell(3,3,b),1)
true(control(white),1)
```

The update rules are basically the same as in GDL. There are just three differences. (1) We add a step argument to does and true. (2) We replace `next` with `true` and add a different step argument. (3) We use `successor` to relate successive steps.

```
true(cell(I,J,x),N) :-
   does(white,mark(I,J),M) &
   successor(M,N)

true(cell(I,J,o),N) :-
   does(black,mark(I,J),M) &
   successor(M,N)

true(cell(K,L,b),N) :-
   does(W,mark(I,J),M) &
   true(cell(K,L,b),M) & distinct(I,K) &
   successor(M,N)

true(cell(K,L,b),N) :-
   does(W,mark(I,J),M) &
   true(cell(K,L,b),M) & distinct(J,L) &
   successor(M,N)

true(cell(I,J,Z),N) :-
   does(W,mark(I,J),M) &
```

```
  true(cell(I,J,Z),M) & distinct(Z,b) &
  successor(M,N)

true(control(white),N) :-
  true(control(black),M) &
  successor(M,N)

true(control(black),N) :-
  true(control(white),M) &
  successor(M,N)
```

Our view relations are the same except that, once again, we add step arguments. In this case, we do not need the successor relation since the same step is involved in all cases.

```
row(M,W,S) :-
  true(cell(M,1,W),S) &
  true(cell(M,2,W),S) &
  true(cell(M,3,W),S)

column(N,W) :-
  true(cell(1,N,W)) &
  true(cell(2,N,W)) &
  true(cell(3,N,W))

diagonal(W) :-
  true(cell(1,1,W)) &
  true(cell(2,2,W)) &
  true(cell(3,3,W))

diagonal(W) :-
  true(cell(1,3,W)) &
  true(cell(2,2,W)) &
  true(cell(3,1,W))

line(W,N) :- row(J,W,N)
line(W,N) :- column(K,W,N)
line(W,N) :- diagonal(W,N)

open(N) :- true(cell(J,K,b),N)
```

Finally, we define the notions of legality, goal, and termination using our augmented versions of these relations.

```
legal(W,mark(X,Y),N) :-
    true(cell(X,Y,b),N) &
    true(control(W),N)

legal(white,noop,N) :-
    true(control(black),N)

legal(black,noop,N) :-
    true(control(white),N)

goal(white,100,N) :- line(x,N) & ~line(o,N)
goal(white,50,N) :- ~line(x,N) & ~line(o,N)
goal(white,0,N) :- ~line(x,N) & line(o,N)

goal(black,100,N) :- ~line(x,N) & line(o,N)
goal(black,50,N) :- ~line(x,N) & ~line(o,N)
goal(black,0,N) :- line(x,N) & ~line(o,N)

terminal(N) :- line(W,N)
terminal(N) :- step(N) & ~open(N)
```

And that's it. Everything is pretty much the same as before. We have just made explicit the Markov character of the game description.

18.4 EXAMPLE–CHESS

In order to see how we can use SDL to encode non-Markov constraints, let's look at the game of Chess and, in particular, the legality conditions for castling and *en passant* captures. While it is possible to describe these conditions in GDL, that description requires that we include some base propositions specifically for this purpose. In SDL, we can describe the conditions directly, without these additional base propositions.

As in our description of Tic-Tac-Toe, we begin with a description of the structural elements of the game—roles, base propositions, percepts (again none), actions, and steps. Here, we have limited the game to 100 steps.

```
role(white)
role(black)
```

```
base(cell(I,J,W)) :-
  file(I) &
  rank(J) &
  filler(W)

base(control(W)) :-
  role(W)

input(W,move(I,J,K,L)) :-
  role(W) &
  file(I) &
  rank(J) &
  file(K) &
  rank(L)

input(W,castleshort) :-
  role(W)

input(W,castlelong) :-
  role(W)

input(W,noop) :-
  role(W)

file(a)
file(b)
file(c)
file(d)
file(e)
file(f)
file(g)
file(h)

rank(1)
rank(2)
rank(3)
rank(4)  .
rank(5)
rank(6)
```

```
rank(7)
rank(8)

filler(wk)
filler(wq)
filler(wb)
filler(wn)
filler(wr)
filler(wp)
filler(bk)
filler(bq)
filler(bb)
filler(bn)
filler(br)
filler(bp)
filler(b)

step(1)
step(2)
   ...
step(99)
step(100)

successor(1,2)
successor(2,3)
   ...
successor(98,99)
successor(99,100)
```

There two types of castling in Chess: castling on the king side (`castleshort`) and castling on the queen side (`castlelong`). In either case, the corresponding rook is moved to the cell adjacent to the king, and the king is placed on the other side of the rook.

Castling is a very useful move in many situations. However, it can only be done if certain conditions are met. First, the king and the rook must be in their original positions and must not have been previously moved. Second, the spaces between the king and the rook must be clear. Third, the spaces from the initial position of the king to the final position must not be under attack.

Let's formalize the first of these conditions in the case of white castling on the king side. (The other conditions pose no special difficulties; and, for the sake of simplicity, we ignore them.)

The corresponding definition of legality for white to castleshort is shown below. The move is legal provided that the king and the rook are on their initial squares and the squares in between are empty and neither the king nor the rook have been moved.

```
legal(white,castleshort,N) :-
   true(cell(e,1,wk),N) &
   true(cell(f,1,b),N) &
   true(cell(g,1,b),N) &
   true(cell(h,1,wr),N) &
   ~moved(wk,N) &
   ~moved(wr,N)
```

The white king has been moved on step N if there was any move from cell e1 on that step or any preceding step. The king-side white rook has been moved on step N if there was any move from cell h1 on that step or any preceding step. Finally, a piece has been moved on a step if it was moved on any preceding step.

```
moved(wk,N) :-
   does(W,move(e,1,K,L),N)

moved(wr,N) :-
   does(W,move(h,1,K,L),N)

moved(P,N) :-
   moved(P,M) &
   successor(M,N)
```

This definition of legality guarantees that, if either the white king or the king-side white rook have been moved, then castling on the king side is not legal. Significantly, the condition is stated using just the state of the board and the history of moves, without any extraneous propositions in the state description.

Now, on to *en passant* captures. If a pawn has not been moved, a player has the option of advancing the pawn either one or two steps forward. If it chooses to advance its pawn two steps, it may move through a cell which is threatened by an opposing pawn. In this case, the opponent has the option of taking the pawn even though it is no longer on a threatened square.

The conditions for *en passant* captures are strict. The player must have moved the pawn for the first time, and it must have made that move on the preceding step. We can formalize these conditions in SDL as shown below. Here, for simplicity, we have shown the conditions for the black player to make the *en passant* capture, and we have shown only the case of a pawn move on the e file and a capture from the f file. (Expanding this for the general case is messy but straightforward.)

```
legal(black,move(f,4,e,3),N) :-
  does(white,move(e,2,e,4),M) &
  successor(M,N)
```

Although, in this example, the rule concerns only two states, it still requires SDL because it defines legality in one state on the basis of a move in the preceding state rather than on conditions in the current state, as in GDL.

This definition guarantees the desired legality conditions; and, as with the preceding example, the condition is stated using just the state of the board and the history of moves, without any extraneous propositions in the state description.

CHAPTER 19

Incomplete Game Descriptions

19.1 INTRODUCTION

In our discussion of General Game Playing thus far, we have assumed that game descriptions are complete. This does not mean that players have full information. For example, in multiple-player games, they do not know the moves of other players before they make those moves; and, in the case of imperfect information games, they do not even know the exact state of the world, only what they can perceive of it. However, in our work thus far, we have assumed that the game rules are complete—they completely define the initial state, legality, percepts, update, goals, and termination in terms of states and actions.

Unfortunately, in real-world settings, complete information is not always available. In some cases, agents do not know all of the effects of their actions; they may not even know exactly which actions are legal or what their rewards are or when a game is over.

Incomplete Game Description Language (or IGDL) is a variant of GDL designed to facilitate the encoding of incomplete game descriptions. In place of rules that define concepts exactly, IGDL has logical sentences that constrain concepts in more or less detail. (Spoiler for those with background in Logic and Logic Programming—the essential difference between GDL and IGDL is that, in IGDL, there is no negation as failure.)

In this chapter, we define IGDL and illustrate its use in writing incomplete descriptions of various sorts. In the next section, we introduce Relational Logic, which is the logical basis for IGDL. In the section after that, we define IGDL in Relational Logic in much the same way that we defined GDL in terms of Logic Programming. We then show how to use IGDL in the context of some examples. Finally, we talk about game management and game play.

19.2 RELATIONAL LOGIC

The basic elements of Relational Logic are the same as those of Logic Programming. We have the same vocabulary—object constants, function constants, relation constants, and variables. We define terms in exactly the same way—as object constants, variables, and functional terms. And we define atomic sentences and literals in the same way as well.

The main difference between the language of Relational Logic and the language of Logic Programming is that, in Relational Logic, we write (1) logical sentences and (2) quantified sentences instead of rules.

There are five types of *logical sentences* in Relational Logic—negations, conjunctions, disjunctions, implications, and equivalences.

A *negation* consists of the negation operator ~ and a simple or compound sentence, called the *target*. For example, given the sentence p(a), we can form the negation of p(a) as shown below.

$$(\sim p(a))$$

A *conjunction* is a sequence of sentences separated by occurrences of the & operator and enclosed in parentheses, as shown below. The constituent sentences are called *conjuncts*. For example, we can form the conjunction of p(a) and q(a,a) as follows.

$$(p(a) \ \& \ q(a,a))$$

A *disjunction* is a sequence of sentences separated by occurrences of the | operator and enclosed in parentheses. The constituent sentences are called *disjuncts*. For example, we can form the disjunction of p(a) and q(a,a) as follows.

$$(p(a) \ | \ q(a,a))$$

An *implication* consists of a pair of sentences separated by the => operator and enclosed in parentheses. The sentence to the left of the operator is called the *antecedent*, and the sentence to the right is called the *consequent*. The implication of p(a) and q(a,a) is shown below.

$$(p(a) \ \Rightarrow \ q(a,a))$$

A *biconditional*, is a combination of an implication and a reverse implication. For example, we can express the biconditional of p(a) and q(a,a) as shown below.

$$(p(a) \ <=> \ q(a,a))$$

Note that the constituent sentences within any compound sentence can be either simple sentences or compound sentences or a mixture of the two. For example, the following is a legal compound sentence.

$$((p(a) \ | \ q(a,a)) \ \Rightarrow \ r(a))$$

A *quantified sentence* in Relational Logic is formed from a *quantifier*, a variable, and an embedded sentence. The embedded sentence is called the *scope* of the quantifier. There are two types of quantified sentences in Relational Logic, viz. universally quantified sentences and existentially quantified sentences.

A *universally quantified sentence* is used to assert that all objects have a certain property. For example, the following expression is a universally quantified sentence asserting that, if p holds of an object, then q holds of that object and itself.

$$(AX:(p(X) \ \Rightarrow \ q(X,X))$$

An *existentially quantified sentence* is used to assert that some object has a certain property. For example, the following expression is an existentially quantified sentence asserting that there is an object that satisfies p and, when paired with itself, satisfies q as well.

$$(EX:(p(X) \& q(X,X))$$

Note that quantified sentences can be nested within other sentences. For example, in the first sentence below, we have quantified sentences inside of a disjunction. In the second sentence, we have a quantified sentence nested inside of another quantified sentence.

$$(AX:p(X)) \mid (EX:q(X,X))$$
$$(AX:(EY:q(X,Y)))$$

One disadvantage of our notation, as written, is that the parentheses tend to build up and need to be matched correctly. It would be nice if we could dispense with parentheses, e.g., simplifying the preceding sentence to the one shown below.

$$p(a) \mid q(a,a) \Rightarrow r(a)$$

Unfortunately, we cannot do without parentheses entirely, since then we would be unable to render certain sentences unambiguously. For example, the sentence shown above could have resulted from dropping parentheses from either of the following sentences.

$$(p(a) \mid q(a,a)) \Rightarrow r(a)$$
$$p(a) \mid (q(a,a) \Rightarrow r(a))$$

The solution to this problem is the use of *operator precedence*. The following table gives a hierarchy of precedences for our operators. The ~ operator has higher precedence than &; & has higher precedence than |; and | has higher precedence than => and <=>.

In unparenthesized sentences, it is often the case that an expression is flanked by operators, one on either side. In interpreting such sentences, the question is whether the expression associates with the operator on its left or the one on its right. We can use precedence to make this determination. In particular, we agree that an operand in such a situation always associates with the operator of higher precedence. When an operand is surrounded by operators of equal precedence, the operand associates to the right.

As with Logic Programming, the *Herbrand base* for Relational Logic language is the set of all ground relational sentences that can be formed from the constants of the language. Said another way, it is the set of all sentences of the form $r(t_1, \ldots, t_n)$, where r is an n-ary relation constant and $t_1, \ldots, t_n$ are ground terms.

A *truth assignment* for a Relational Logic language is a function that maps each ground relational sentence in its Herbrand base to a truth value. For example, the truth assignment defined below is an example for the case of the language with object constants and b, unary relation constant p, and binary relation constant q.

$$
\begin{aligned}
p(a) &\to 1 \\
p(b) &\to 0 \\
q(a,a) &\to 1 \\
q(a,b) &\to 0 \\
q(b,a) &\to 1 \\
q(b,b) &\to 0
\end{aligned}
$$

Once we have a truth assignment for the ground relational sentences of a language, the semantics of our operators prescribes a unique extension of that assignment to the complex sentences of the language. A truth assignment satisfies a negation $\sim\phi$ if and only if it does not satisfy ϕ. A truth assignment satisfies a conjunction $(\phi_1 \& \dots \& \phi_n)$ if and only if it satisfies every ϕ_i. A truth assignment satisfies a disjunction $(\phi_1 \mid \dots \mid \phi_n)$ if and only if it satisfies at least one ϕ_i. A truth assignment satisfies an implication $(\phi \Rightarrow \psi)$ if and only if it does not satisfy ϕ or does satisfy ψ. A truth assignment satisfies an equivalence $(\phi \Leftrightarrow \psi)$ if and only if it satisfies both ϕ and ψ or it satisfies neither ϕ nor ψ.

In order to define satisfaction of quantified sentences, we need the notion of instances. An *instance* of an expression is an expression in which all variables have been consistently replaced by ground terms. Consistent replacement here means that, if one occurrence of a variable is replaced by a ground term, then all occurrences of that variable are replaced by the same ground term.

A universally quantified sentence is true for a truth assignment if and only if *every* instance of the scope of the quantified sentence is true for that assignment. An existentially quantified sentence is true for a truth assignment if and only if *some* instance of the scope of the quantified sentence is true for that assignment.

As an example of these definitions, consider the sentence AX:(p(X) => q(X,X)). What is the truth value under the truth assignment shown above? According to our definition, a universally quantified sentence is true if and only if every instance of its scope is true. For this language, with object constants a and b and no function constants, there are just two instances. See below.

$$
\begin{aligned}
p(a) &\Rightarrow q(a,a) \\
p(b) &\Rightarrow q(b,b)
\end{aligned}
$$

We know that p(a) is true and q(a,a) is true, so the first instance is true. q(b,b) is false, but so is p(b) so the second instance is true as well. Since both instances are true, the original quantified sentence is true.

Now let's consider a case with nested quantifiers. Is AX:EY:q(X,Y) true or false for the truth assignment shown above? As before, we know that this sentence is true if every instance of its scope is true. The two possible instances are shown below.

$$
\begin{aligned}
&\text{EY:q(a,Y)} \\
&\text{EY:q(b,Y)}
\end{aligned}
$$

To determine the truth of the first of these existential sentences, we must find at least one instance of the scope that is true. The possibilities are shown below. Of these, the first is true; and so the first existential sentence is true.

$$q(a,a)$$
$$q(a,b)$$

Now, we do the same for the second existentially quantified. The possible instances follow. Of these, again the first is true; and so the second existential sentence is true.

$$q(b,a)$$
$$q(b,b)$$

Since both existential sentences are true, the original universally quantified sentence must be true as well.

We say that a truth assignment *satisfies* a sentence with free variables if and only if it satisfies every instance of that sentence. A truth assignment *satisfies* a set of sentences if and only if it satisfies every sentence in the set.

19.3 INCOMPLETE GAME DESCRIPTION LANGUAGE

IGDL is not so much a language as a family of languages. It has multiple dialects - one for each of the various dialects of GDL. There is IGDL corresponding to GDL; there is IGDL-II, corresponding to GDL-II; and there is ISDL, corresponding to SDL.

Given one of these dialects of GDL, the IGDL variant is obtained using the language of Relational Logic in place of the language of Logic Programming. In other words, instead of writing rules, one writes logical sentences. Everything else remains the same.

We make this more concrete in the following sections by looking at various descriptions of a single game. First, we look at Buttons and Lights written in GDL. We then look at a complete description of the game written in IGDL-II. And then we look at an incomplete description written in IGDL-II. Finally, we discuss game management and game play with incomplete descriptions.

19.4 BUTTONS AND LIGHTS REVISITED

In this section, we return to the game of Buttons and Lights. Recall that, in ordinary Buttons and Lights, there are three base propositions (the lights) and three actions (the buttons). See below. Pushing the first button in each group toggles the first light; pushing the second button in each group interchanges the first and second lights; and pushing the third button in each group interchanges the second and third lights. Initially, the lights are all off. The goal is to turn on all of the lights. The game terminates on step 7 (after six moves).

The ordinary GDL for this game is shown below. There is just one role, here called robot. There are three base propositions, one percept, three actions, and seven steps (with the usual

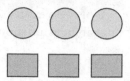

Figure 19.1: Buttons and lights.

successor relation). The robot can see proposition q whenever it is true. All three actions are legal in all states. The update rules, goal, and termination are as just described.

```
role(robot)

base(p)
base(q)
base(r)

percept(robot,q)

input(robot,a)
input(robot,b)
input(robot,c)

step(1)
step(2)
step(3)
step(4)
step(5)
step(6)
step(7)

successor(1,2)
successor(2,3)
successor(3,4)
successor(4,5)
successor(5,6)
successor(6,7)

sees(q) :- true(q)
```

```
legal(robot,a)
legal(robot,b)
legal(robot,c)

next(p) :- does(robot,a) & ~true(p)
next(p) :- does(robot,b) & true(q)
next(p) :- does(robot,c) & true(p)
next(q) :- does(robot,a) & true(q)
next(q) :- does(robot,b) & true(p)
next(q) :- does(robot,c) & true(r)
next(r) :- does(robot,a) & true(r)
next(r) :- does(robot,b) & true(r)
next(r) :- does(robot,c) & true(q)

goal(robot,100) :- true(p) & true(q) & true(r)
goal(robot,0) :- ~true(p)
goal(robot,0) :- ~true(q)
goal(robot,0) :- ~true(r)

terminal :- true(7)
```

In playing games with incomplete descriptions, the game manager starts with a complete description, typically encoded in GDL, like the one shown here. However, the players are given only partial descriptions. These partial descriptions are written in IGDL or IGDL-II or ISDL. In the next section, we look at a complete description of Buttons and Lights written in IGDL-II; and in the section after that we look at an incomplete description.

19.5   COMPLETE DESCRIPTION OF BUTTONS AND LIGHTS

The first step in writing an IGDL-II description is the same as for GDL, GDL-II, or SDL: we enumerate the structural components of the game—roles, base propositions, percepts, actions, and steps. In almost all cases, these descriptions are complete, as in this case.

The IGDL-II description of these structural components is shown below. The main difference here is that we have written negative sentences for each relation to tell us what is not true. In Logic Programming, this is not necessary, since anything that is not known to be true there is assumed to be false. In incomplete descriptions, we cannot make this assumption. If we do not know something, it is not necessarily false; we just do not know whether it is true or false. So, we need to be explicit about the things we know to be false.

```
role(robot)
X!=robot => ~role(X)

base(p)
base(q)
base(r)
X!=p & X!=q & X!=r => ~base(X)

percept(robot,q)
X!=robot | Y!=q => ~percept(X,Y)

input(robot,a)
input(robot,b)
input(robot,c)
X!=robot | Y!=a & Y!=b & Y!=c => ~input(X,Y)

step(1)
step(2)
step(3)
step(4)
step(5)
step(6)
step(7)
X!=1 & X!=2 & X!=3 & X!=4 & X!=5 & X!=6 & X!=7 => ~step(X)

successor(1,2)
successor(2,3)
successor(3,4)
successor(4,5)
successor(5,6)
successor(6,7)
(X!=1 | Y!2) & ... & (X!=6 | Y!7) => ~successor(X,Y)
```

The other components of the game description can be formalized in the same way. See below. All lights are off in the initial state. The player can see its single percept. All three actions are legal. The update rules are the same. And the goal and termination rules are the same.

```
~init(X)

true(q) <=> sees(q)
X!=q => ~sees(X)

legal(robot,a)
input(robot,b)
input(robot,c)
X!=robot | Y!=a & Y!=b & Y!=c => ~legal(X,Y)

next(p) <=>
  does(robot,a) & ~true(p) |
  does(robot,b) & true(q) |
  does(robot,c) & true(p)

next(q) <=>
  does(robot,a) & true(q) |
  does(robot,b) & true(p) |
  does(robot,c) & true(r)

next(r) <=>
  does(robot,a) & true(r) |
  does(robot,b) & true(r) |
  does(robot,c) & true(q)

X!=p & X!=q & X!=r => ~next(X)

goal(robot,100) <=> true(p) & true(q) & true(r)
goal(robot,0) <=> ~true(p) | ~true(q) | ~true(r)
X!=robot | (Y!100 & Y!=0) => ~goal(X,Y)

terminal <=> true(7)
```

The sentences here are a little different from those in the GDL description. However, with a little reflection on the semantics of GDL-II and IGDL-II, it is easy to see that they describe exactly the same game.

19.6   INCOMPLETE DESCRIPTION OF BUTTONS AND LIGHTS

As a simple example of an *incomplete* description written in IGDL-II, consider a simple variation of Buttons and Lights in which the players know everything except the initial state of the game.

To be more precise, we take all of the sentences from Section 19.5, with the exception of the sentence asserting that none of the base propositions are true in the initial state. Removing this sentence means that the player starts the game in any one of eight possible states.

Of course, once the player gets a percept, i.e., it sees whether q is true or false, this ambiguity is cut down to just four states. Unfortunately, in this situation there is still no guaranteed solution in the allowed number of steps. There are just too many cases to consider.

We can make things a little better by giving the player a bit more information. Let's say we tell the player that proposition p and proposition q have the same initial value. This can be done by augmenting the description with a sentence like the one shown below.

```
init(p) <=> init(q)
```

With this additional information, the player knows that the game starts in one of four possible states. Once it is given its initial percept, it can cut this down to just two initial states. And, by clever play, it can then solve the problem despite the residual ambiguity, as described in the next section.

19.7   PLAYING BUTTONS AND LIGHTS WITH AN INCOMPLETE DESCRIPTION

Game management and play with incomplete descriptions is a little different from management and play with complete information.

First of all, the manager and the players typically have different descriptions of the game. The manager has a complete description, like the one in Sections 19.4 and 19.5. This is necessary so that it can simulate the game accurately. However, the players have only partial descriptions, like the one in Section 19.6. In some cases, it is even possible for different players to have different descriptions.

Second, our usual techniques for game play do not necessarily work. For example, with incomplete descriptions, the players may not know the initial state exactly (as is the case in the description of Section 19.6). Or they may know the initial state but not be able to determine a unique successor state, given limited information about the update rules for the game. They may not even know in all cases whether the game is over.

Dealing with limitations of these sorts means that players must keep open multiple options on state (as in playing games in IGDL-II). Also, they must use the description in new and interesting ways to extract as much information from the description as possible, possibly combining from multiple time steps.

To illustrate these ideas, let's consider the sort of computation necessary in the context of the incomplete game description given in Section 19.6.

When the game begins, the player is given its percept and learns that q is false. From this and the partial constraint on init, it knows that p must also be false. It still does know whether r is true or false, so there are two possible states to consider. See below.

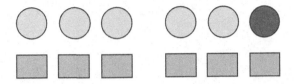

Figure 19.2: Possible states of Buttons and Lights on step 1.

There are several ways to proceed in this situation. In one approach, the player presses the a button to make p true, and then it presses the b button to interchange p and q. This makes p false and q true.

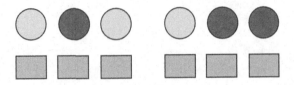

Figure 19.3: Possible states of Buttons and Lights on step 3.

On the third step, it presses the c button to interchange q and r. This makes r true and gives q the value of r.

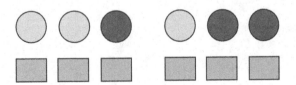

Figure 19.4: Possible states of Buttons and Lights on step 4.

Now, if q is false after this operation, then the player knows that r was false in the initial state. In this case, it needs to press a again, followed by b, followed by a to get all three lights on.

If q is true after interchanging with r, then the player just needs to press a to make p true, at which point all three lights are on. It can then press b or c a couple of times to get to step 7.

The challenge in building players for IGDL descriptions is making them able to use constraints that can help it to infer as much information about the state as possible from their perceptual inputs and the known constraints. In some cases, like this one, the job is easy. In other cases, especially with ISDL, the constraints may involve reasoning about multiple time steps, and the process can be extremely complex.

CHAPTER 20

Advanced General Game Playing

20.1 INTRODUCTION

Although we have looked at a few variations on traditional GGP, the variations have been relatively minor. Over the years, researchers have proposed more extreme variants. In this chapter, we briefly describe some of the more popular proposals, viz. Temporal General Game Playing, Inductive General Game Playing, Really General Game Playing, and finally Enhanced General Game Playing.

20.2 TEMPORAL GENERAL GAME PLAYING

In traditional General Game Playing, players are given two clocks—one for initialization and one for game play—and they are both hard limits. Over the years, members of the community have proposed other sorts of timing restrictions.

One possibility is to replace the hard time limits on each step with a cumulative clock, with the idea that each player stops its clock as soon as it replies to the game manager's play request. This way, players can conserve their time for portions of a match requiring more attention. The downside of cumulative time is that players must decide for themselves how to allocate their time; and this can add to the difficulty of building effective systems.

A different approach is to connect playing time with reward. For example, a player might get a reward that depends not just on the game state but also on the amount of time spent computing its moves. Again, this is a complication, but it allows us to model some real-world applications that do not fit the fixed clock or cumulative time models.

Finally, some people have proposed making the world dynamic, so that the world state changes while the players are contemplating their moves, placing emphasis on economy of deliberation and careful timing of action.

20.3 INDUCTIVE GENERAL GAME PLAYING

In traditional General Game Playing, the players are given game descriptions at runtime. Usually, these game descriptions are complete; and, even then, the task of playing such games is difficult. As we have seen, in some cases, the descriptions are incomplete; and this complicates the process of playing games. Inductive General Game is a variant of General Game Playing that is even more difficult.

In Inductive General Game Playing (IGGP), the players are given no rules at all. In the place of rules, they are provided with a corpus of records for game matches. Given only this corpus of data, they must figure out the rules of the game for themselves; and then, in traditional GGP fashion, they must use these rules to play the games effectively.

One thing that makes the job a little easier is that there is no noise, i.e., no errors in the match records. All matches are correctly played games. At the same time, the job is complicated by the fact that there are no negative examples, i.e., match records that do not correspond to correctly played games. Of course, both of these limitations can easily be rectified. It is simple to add in some false positives and no problem at all in supplying negative examples as well, with or without false negatives.

20.4   REALLY GENERAL GAME PLAYING

Really General Game Playing (RGGP) extends this progression of difficulty one step further. In RGGP, players are given structural information about their games and a simple utility sensor but that is all. They know the roles, the percepts, and the actions of each player; but that is all. They are not even given samples of games.

In RGGP, players explore the world, reading their utility sensors. They must then develop theories of how the world works and use those theories to optimize the readings of their utility sensors.

20.5   ENHANCED GENERAL GAME PLAYING

Finally, some have proposed a variant of GGP with *more* information rather than less. This is often called Enhanced General Game Playing (EGGP).

The extra information in EGGP might included identity information about their opponents. With this information, players can do meaningful opponent modeling and re-use that analysis from one match to another.

The extra information in EGGP might also include tournament information, e.g., whether it is a cumulative score tournament, a single elimination ladder, a double elimination ladder, and so forth, so that it can strategize about how to proceed. For example, in a cumulative score tournament, a player should try to maximize its return on every game, whereas in an elimination ladder, all it needs to do is to beat its opponent.

In its most general form, EGGP is interesting because a player's decisions do not end when a match is over. It is playing a bigger game, which can involve the multiple matches of a tournament and even the results of multiple tournaments. For an EGGP player of this sort, its entire existence constitutes one big game.

Authors' Biographies

MICHAEL GENESERETH

Michael Genesereth is an associate professor in the Computer Science Department at Stanford University. He received his Sc.B. in Physics from M.I.T. and his Ph.D. in Applied Mathematics from Harvard University. Prof. Genesereth is most known for his work on computational logic and applications of that work in enterprise computing, computational law, and general game playing. He is the current director of the Logic Group at Stanford and founder and research director of CodeX (The Stanford Center for Legal Informatics). He initiated the International General Game Playing Competition in 2005.

MICHAEL THIELSCHER

Michael Thielscher is a professor in the School of Computer Science at the University of New South Wales Australia. He received his postgraduate diploma, Ph.D., and Higher Doctorate in Computer Science from Darmstadt University in Germany. Prof. Thielscher is most known for his work in knowledge representation, cognitive agents and robots, and general game playing. He is a current associate director of the iCinema Center for Interactive Cinema Research at UNSW and an adjunct professor at the University of Western Sydney. He has published numerous papers on general game playing and led the development of FLUXPLAYER, a previous winner of the International General Game Playing Competition.

Printed in the United States
by Baker & Taylor Publisher Services